Springer Tracts in Modern Physics

Springer Tracts in Modern Physics provides comprehensive and critical reviews of topics of current interest in physics. The following fields are emphasized: elementary particle physics, solid-state physics, complex systems, and fundamental astrophysics.
Suitable reviews of other fields can also be accepted. The editors encourage prospective authors to correspond with them in advance of submitting an article. For reviews of topics belonging to the above mentioned fields, they should address the responsible editor, otherwise the managing editor.
See also http://www.springer.de/phys/books/stmp.html

Springer Tracts in Modern Physics
Volume 185

Now also Available Online

Starting with Volume 165, Springer Tracts in Modern Physics is part of the Springer LINK s
For all customers with standing orders for Springer Tracts in Modern Physics we offer t
text in electronic form via LINK free of charge. Please contact your librarian who can re
password for free access to the full articles by registration at:

http://link.springer.de/series/stmp/reg_form.htm

If you do not have a standing order you can nevertheless browse through the table of cont
the volumes and the abstracts of each article at:

http://link.springer.de/series/stmp/

There you will also find more information about the series.

Springer
Berlin
Heidelberg
New York
Hong Kong
London
Milan
Paris
Tokyo

Volkmar Dierolf

Electronic Defect States in Alkali Halides

Effects of Interaction with Molecular Ions

With 80 Figures and 22 Tables

Dr. Volkmar Dierolf

Lehigh University
Physics Department
16 Memorial Drive East
Bethlehem, PA 18015, USA
E-mail: vod2@lehigh.edu

Library of Congress Cataloging-in-Publication Data.
Cataloging-in-Publication Data applied for

A catalog record for this book is available from the Library of Congress.

Bibliographic information published by Die Deutsche Bibliothek.
Die Deutsche Bibliothek lists this publication in the Deutsche Nationalbibliografie; detailed bibliographic data is available in the Internet at http://dnb.ddb.de.

Physics and Astronomy Classification Scheme (PACS):
61.72Bb, 61.72Ji, 63.20Pw, 73.30.-v, 78.55Fv

ISSN print edition: 0081-3869
ISSN electronic edition: 1615-0430
ISBN 3-540-00471-8 Springer-Verlag Berlin Heidelberg New York

Springer-Verlag Berlin Heidelberg New York
a member of BertelsmannSpringer Science+Business Media GmbH

http://www.springer.de

Printed in Germany

Typesetting: Author and LE-TeX GbR, Leipzig using a Springer LaTeX macro package
Cover concept: eStudio Calamar Steinen
Cover production: *design & production* GmbH, Heidelberg

Printed on acid-free paper 56/3141/YL 5 4 3 2 1 0

In honor of Professor F. Luty;

dedicated to Mihoko

Preface

The initiator of the field of electronic–molecular defect complexes and consequently also the basis of this work was Professor F. Luty of the University of Utah. In a Physical Review Letter of 1983 he reported together with his colleague and coworker Y. Yang, two *first time observations*:

- E–V energy transfer from an electronic F center defect to the stretch-mode vibration of a neighboring molecular ion (CN^-)
- vibrational luminescence of a molecule in an ionic solid with almost completely radiative transition rates.

With these findings, they showed that it is still possible to discover unexpected, exciting physics even in such "simple" and "well understood" systems as substitutional defects in alkali halides. From early on Professor W. von der Osten of the University of Paderborn, who was on sabbatical in Utah in 1984, and Professor J.-M. Spaeth, also of the University of Paderborn became involved, with their respective research groups and experimental specialities, and made significant contributions to the field. I was very lucky to have the unique opportunity to work in the groups of these major players in the field and to profit from their experience for which I am very grateful.

In this text I attempt to summarize the current state of investigations in this field to which many more research groups have now contributed, but will focus on the progress which has been achieved since the last review was published by Luty and his coworkers in 1993. Although I have tried to include as many aspects as possible, this will naturally be a somewhat subjective summary centered around the work in which I was directly involved. For that reason, I would like to ask for understanding from anyone who may think that his/her contribution has not been covered adequately. The text includes a large amount of previously unpublished data, which is the product of a collaboration with Professor F. Luty and Dr. C.P. An at the University of Utah (as indicated in the respective chapters). Please give proper credit to them when quoting these results. Last but not least, I would like to thank E. Gustin, J. Hoidis, M. Leblans, T. Pawlik, U. Rogulis, D. Samiec, and M. Yoshida-Dierolf for experimental support and many fruitful discussions.

Lehigh University, July 2002 *Volkmar Dierolf*

Contents

1 Introduction and Historical Overview

F centers, monoatomic defects, and molecular impurities in alkali halides are generally considered as "classic fields" of solid-state physics, which have been studied systematically since the early decades of the 20th century. It is a common belief that essentially everything is understood. While the latter statement is certainly not true even for individual defects, as becomes apparent as soon as one looks in somewhat more detail, it is quite surprising that investigations of combinations of these defects, i.e. complexes combining electronic and atomic defects[1] with molecular ions, were only started in the early 1980s.

The course of the investigations within this field was by no means straight, and exhibited some features typical of "good" research in physics:

- It started with a very good idea based on intuition, which, however, failed at first.
- Nevertheless, the field took off because very unexpected and exciting new results and properties were discovered.
- In the end, the initial idea and expectation were fulfilled for a defect system of the same kind as but different from the one which was studied initially. Ironically, this system was studied only to investigate the newly found features of the systems in question.

1.1 The Initial Idea

Color center lasers, which were first demonstrated by Fritz and Menke [1] in 1965 and, after the availability of powerful laser pump sources, by Mollenauer and Olson [2] in 1974, were promising laser candidates in the late 1970s and 1980s owing to their wide individual tunability range and variability of the central wavelength, which now spans a range from 800 nm to 4 μm with the use of different host materials and defect systems [3]. All lasing systems are based on some kind of aggregated center type (F_A, F_B, F_2^+, etc.); these tend to dissociate at room temperature so that these lasers in most cases have to

[1] Because we are considering in the following only the electronic states of the F centers and the atomic defects, we refer to both types of defects as "electronic" defects.

be kept at low temperature at all times, thereby drastically restricting their usefulness outside of the research laboratory. The only exception is laser systems with a LiF host material. The simple F center, which would not have that disadvantage, has been shown not to be a potential candidate, owing to a combination of a long radiative lifetime and a nonradiative decay channel for higher F center concentrations. While the latter is inevitable owing to the large interaction cross section of the F center's relaxed, excited state (RES), the former is at least partially due to the dominant 2s-like character of the RES, which makes the emission to the 1s ground state parity-forbidden. It has been shown by Bogan and Fitchen [4], however, that the application of an external electric field will mix the RES with the close-lying 2p state, enhancing the radiative transition rate. A short enough radiative lifetime would, in principle, allow laser operation for F center concentrations sufficiently low to make the nonradiative channel negligible. A simple but clever idea (first put forward by Luty) was to replace the external field by an internal one, which can be produced by the electric dipole of a molecular defect, thereby offering an intrinsic method to enhance the radiative transition probability. As a prime candidate, the OH^- molecule, which can easily (often unintentionally) be doped into most alkali halide crystals was chosen.

The result of the experiments, however, was a "complete disaster": above a critical OH^- concentration the F luminescence, instead of being enhanced, was completely quenched [5].

1.2 The Unexpected New Results

Despite this failure, the investigations were extended to CN^- molecules. This brave decision was richly rewarded by the discovery of two phenomena which until then had been believed to be insignificant, because they contradicted well-established "rules of thumb" [6, 7]:

- E–V energy transfer from an electronic F center defect to stretch-mode vibration of a neighboring molecular ion (CN^-). In Cs halides this energy transfer is so efficient that the electronic F center emission disappears almost completely. At present this high efficiency cannot be described quantitatively.
- Vibrational luminescence (VL) of a molecule in an ionic solid with almost completely radiative transition rates.

Both phenomena become apparent in a single experiment, which will be described in more detail later on: optical excitation of the F center/CN^- complex in the range of its electronic F-center-like transitions in the visible leads to a strong vibrational luminescence in the mid-infrared (MIR) around 5 μm (see Fig. 1.1). The evenly spaced emission lines indicate that the molecule is excited up to its fifth vibrational state, from which it decays stepwise down the vibrational ladder. Until its first observation, vibrational luminescence was believed not to be possible in solids owing to the long radiative lifetime of the vibrational transitions, which should give the system plenty of time to relax nonradiatively into libration/rotational modes or phonons. However, the diatomic CN^- molecule turns out to be a favorable case because it lacks other internal modes that are present for polyatomic molecules such as ReO_4^-, studied intensively by Sievers and coworkers [8, 9], and its rotational potential barriers are fairly low, such that many librational/rotational quanta are necessary to dissipate the excitation energy.

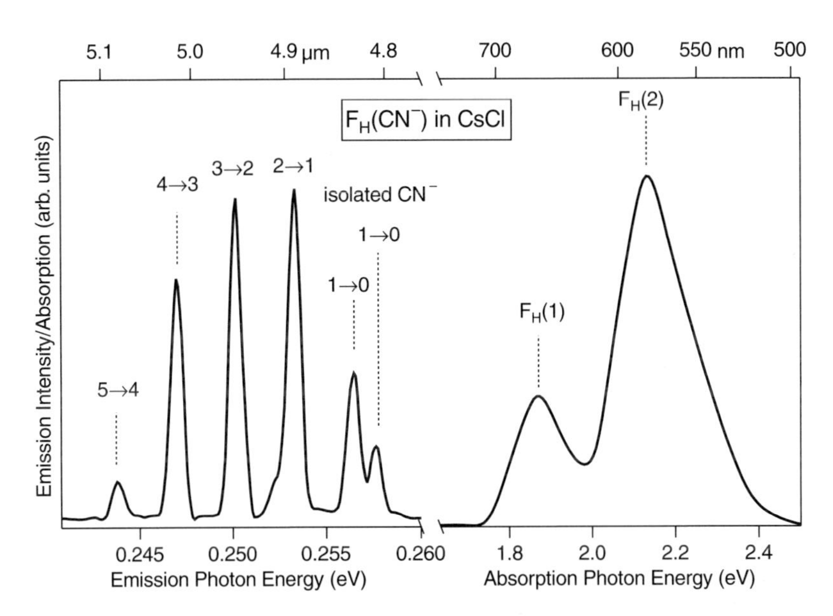

Fig. 1.1. Absorption and vibrational luminescence of F center–CN^- complexes in CsCl

Shortly after the spontaneous luminescence of CN^- molecules had been demonstrated, these molecules were used for the first demonstrations of a vibrational laser in an ionic solid [10]. This was achieved by exciting the molecules directly using a frequency-doubled CO_2 laser. In parallel with those efforts and much more convincingly, laser action was demonstrated employing an indirect pumping scheme via very effective E–V transfer using just a simple He–Ne laser as a pump source [11, 12]. Although these lasers have very promising properties (low threshold and high quantum efficiency) a widespread application of these vibrational lasers is prevented by the low temperatures ($T < 60$ K)required for their operation.

1.3 The Dream Comes True

In none of the F center systems studied could the initial hope of decreasing the radiative lifetime by means of molecular defects be fulfilled. Nevertheless the effort put into codoping samples with molecular and electronic defects has delivered the desired improvement in two other ways.

- The radiative transition rate of the symmetry-forbidden Yb^{2+} emission is drastically reduced by the presence of CN^- molecular neighbors. This effect is most pronounced if the CN^- molecular defect is excited optically or thermally making the Yb^{2+}–CN^- complex a potential candidate as the active element in phosphors and lasers.
- A whole class of color center lasers (e.g.: $(F_2^+)_{AH}$) is based on F_2^+ centers stabilized by O_2^- defects produced in OH^--doped crystals through electron irradiation [3]. As the role of the OH^- ion is only indirect, the $(F_2^+)_{AH}$ laser will not be treated in this text.

1.4 What to Expect?

The choice of the defect systems which will be discussed in this text was guided primarily not by possible applications but by a search for "simple" model systems which could be characterized and understood in good depth and could provide a basis for new investigations and for explanation of phenomena that have already been observed in more complex systems but are not sufficiently understood. For these reasons, the simplest possible defect combinations and host materials have been picked:

- Alkali halide hosts with a simple cubic lattice (Cs halides) or a face-centered cubic (Na, K, and Rb-halides and CsF) and a wide bandgap suitable for optical investigations of defects with transitions ranging from the IR to the UV.
- Simple and individually well-understood electronic or monoatomic defects such as F centers, rare-earth ions and ions with the ns^2 ground state configuration. For all those defects, specific electronic transitions can be observed in the bandgaps of the alkali halides. As we are interested only in the electronic transitions of those defects, we refer to this defect class collectively as the "electronic" defect partner.
- Diatomic molecular defects (OH^- and CN^-) with only a single internal vibrational mode. In addition to this mode these molecular defects differ from monoatomic defects by their rotational behavior, which has been studied and is well understood in terms of the barrier heights and energy minima of the cubic-lattice potential, which hinders free rotation and leads to tunneling or classical reorientation between equivalent center arrangements [13, 14].

For the chosen systems, it is always possible to study the mutual static and dynamic interaction effects from both the electronic-defect and the molecular-defect side and therefore a quite complete characterization of the defect complexes can be achieved. Although the main intention of this text is to review the current state of the investigations of these systems and their relevance to other fields, the amount of previously unpublished new data makes it necessary to present a considerable amount of detail in several chapters.

As will be shown, it is not possible so far to give a complete unified theoretical description of all the observed phenomena so therefore only the basic well-established theoretical treatments will be reviewed to begin with (Chaps. 3 and 4). We shall find, however, that in some cases these models will come up against their limitations will not be able to describe all observations, and therefore will have to be extended. For that purpose phenomenological and more system-specific models have to be introduced. This will be done along with presentation of the experimental findings. On the basis of these considerations, the following structure has been chosen:

- The basic properties of the individual defects relevant in our context.
- Theoretical models for the description of molecular defects and E–V transfer.
- Introduction, experimental results, and their interpretation for different defect systems.
- A summarizing discussion and a consideration of the outlook.

Owing to the nature of the defect complexes, a great variety of spectroscopic techniques has been applied, making use of the particular feature of these complexes that the effects of interaction can be investigated from the side of the molecular defect in the IR spectral region and also from the side of the electronic defect in the visible to UV. To give an overview, the most important experimental techniques employed in these studies of the defect systems are listed in the following:

- Electronic absorption (EA) and luminescence (EL) studies in the visible spectral region [6, 15, 16, 17, 18, 19, 20].
- Vibrational absorption (VA) and luminescence (VL) spectroscopy [6, 7, 15, 17, 20, 21, 22, 23].
- Anti-Stokes resonant Raman scattering (ASRR) [24, 25, 26, 27, 28, 29, 30, 31, 32, 33].
- Transient absoprtion (TA) and excited-state absorption [34, 35, 36].
- Magnetic circular dichroism of absorption (MCDA) [37, 38].
- Magnetic resonance studies, both conventionally detected (EPR) and optically detected (ODESR and ODENDOR) [39, 40, 41, 42, 43, 44].
- Two-photon absorption [45].

- Combined excitation/emission spectroscopy (CEES)[2] [51].
- Time-dependent measurements of the ground state recovery [34, 52, 53].

Most techniques are standard and/or are described in detail in the references cited.

Several physical phenomena have been studied by use of these experimental techniques:

Table 1.1. List of commonly used abbreviations

APES	adiabatic potential energy surface
ASR	anti-Stokes Raman scattering
ASRR	anti-Stokes resonance Raman scattering
CC	configurational coordinate (diagram)
CEES	combined excitation-emission spectroscopy
EA	electronic absorption
E–E	(energy transfer) between electronic defects
EL	electronic luminescence
ENDOR	electron–nuclear double resonance
EPR	electron paramagnetic resonance
E-V	(energy transfer) from electronic to vibrational defects
FC	Franck–Condon (factor)
FD	Förster–Dexter (model)
GS	ground state
IR	infrared (spectral region) $\lambda > 700$ nm
MCDA	magnetic circular dichroism of absorption
MIR	mid-IR ($2 < \lambda < 10$ µm, range of molecular stretchmodes)
NIR	near IR (700 nm $< \lambda < 2$ µm)
ODENDOR	optically-detected ENDOR
ODESR	optically-detected electron spin resonance
RES	relaxed excited state
S	Huang–Rhys factor
T_1	temperature at which photoionization of F centers is possible
T_2	aggregation temperature for F-center-related defects
TA	transient absorption
URES	unrelaxed excited state
UV	ultraviolet spectral region ($\lambda < 400$ nm)
VA	vibrational absorption
V–E	(energy transfer) from vibrational to electronic defects
VIS	visible spectral region (400 nm $< \lambda <$ 700 nm)
VL	vibrational luminescence
V–V	(energy transfer) between vibrational defects

[2] A detailed description of the CEES method, including several specific applications (Stark effect measurements and double-resonance excitation–emission spectroscopy in $LiNbO_3$ waveguides) can be found in several Diploma theses [46, 47, 48] and publications [49, 50].

- Shifts of transition energies.
- Optically induced bistability.
- Changes in luminescence properties.
- E–V energy transfer.
- Dynamic coupling effects.

Although most of these phenomena can be observed for all systems, they will be discussed in most detail for those defect systems for which the particular effect is most pronounced. Moreover, special focus will given to the relationships between these phenomena.

Within the field of molecular and electronic defects many abbreviations are used, which may be familiar to experts but will be new to others. For that reason, the abbreviations used within this text are collected in Table 1.1.

References

1. B. Fritz and E. Menke, Solid State Commun. **3**, 61 (1965).
2. L. Mollenauer and D. Olson, Appl. Phys. Lett. **24**, 386 (1974).
3. W. Gellermann, J. Phys. Chem. Solids **52**, 249 (1991).
4. L. D. Bogan and D. Fitchen, Phys. Rev. B **1**, 4122 (1970).
5. L. Gomes and F. Luty, Phys. Rev. B **30**, 7194 (1984).
6. Y. Yang and F. Luty, Phys. Rev. Lett. **51**, 419 (1983).
7. Y. Yang, W. von der Osten, and F. Luty, Phys. Rev. B **32**, 2724 (1985).
8. W. E. Moerner, A. R. Chraplyvy, and A. J. Sievers, Phys. Rev. Lett. **47**, 1082 (1981).
9. W. E. Moerner, A. R. Chraplyvy, and A. J. Sievers, Phys. Rev. B **29**, 6694 (1984).
10. T. R. Gosnell, A. J. Sievers, and C. R. Pollock, Opt. Lett. **10**, 125 (1985).
11. W. Gellermann, Y. Yang, and F. Luty, Opt. Commun. **57**, 196 (1986).
12. W. Gellermann and F. Luty, Optics Comm. **72**, 214 (1987).
13. F. Luty, J. Phys. (Paris) Coll. **C-4**, 120 (1967).
14. F. Luty, in *Defects in Insulating Crystals*, edited by V. Turkevich and K. Shvarts (Springer, Berlin, 1981), pp. 69–89.
15. M. Krantz and F. Luty, Phys. Rev. B **37**, 8412 (1988).
16. G. Baldacchini, S. Botti, U. M. Grassano, L. Gomes, and F. Luty, Europhys. Lett. **9**, 735 (1989).
17. A. Naber, in *Defects in Insulating Materials*, edited by O. Kanert and J.-M. Spaeth (World Scientific, Singapore, 1993), p. 543.
18. J. West, K. T. Tsen, and S. H. Lin, Phys. Rev. B **50**, 9759 (1994).
19. J. West, K. T. Tsen, and S. H. Lin, Mod. Phys. Lett. B **9**, 1759 (1995).
20. V. Dierolf and F. Luty, Phys. Rev. B **54**, 6952 (1996).
21. Y. Yang and F. Luty, J. Lumin. **40&41**, 565 (1988).
22. F. Luty and V. Dierolf, in *Defects in Insulating Materials*, edited by O. Kanert and J.-M. Spaeth (World Scientific, Singapore, 1993), p. 17.
23. A. Naber, Ph.D. thesis, Westfälische Wilhelms-Universität Münster, 1993.
24. K. T. Tsen, G. Halama, and F. Luty, Phys. Rev. B **36**, 9247 (1987).

25. F. Rong, Y. Yang, and F. Luty, Cryst. Latt. Defects Amorph. Mater **18**, 1 (1989).
26. G. Halama, K. T. Tsen, S. H. Lin, F. Luty, and J. B. Page, Phys. Rev. B **39**, 13457 (1989).
27. G. Halama, S. H. Lin, K. T. Tsen, F. Luty, and J. B. Page, Phys. Rev. B **41**, 3136 (1990).
28. G. Halama, K. T. Tsen, S. H. Lin, and J. B. Page, Phys. Rev. B **44**, 2040 (1991).
29. G. Cachei, H. Stolz, W. von der Osten, and F. Luty, J. Phys.: Condens. Matter **1**, 3239 (1989).
30. R. Albrecht, H. Stolz, and W. von der Osten, J. Phys.: Condens. Matter **4**, 9269 (1992).
31. E. Gustin, Ph.D. thesis, Universitaire Instelling Antwerpen (UIA), 1995.
32. E. Gustin, M. Leblans, A. Bouwen, and D. Schoemaker, Phys. Rev. B **54**, 6963 (1996).
33. V. Dierolf, E. Gustin, D. Schoemaker, and F. Luty, J. Lumin. **76&77**, 526 (1998).
34. D. Samiec, H. Stolz, and W. von der Osten, Phys. Rev. B **53**, 8811 (1996).
35. V. Dierolf, J. Hoidis, D. Samiec, and W. von der Osten, J. Lumin. **76&77**, 581 (1998).
36. V. Dierolf, J. Hoidis, D. Samiec, and W. von der Osten, Radiat. Eff. Defects Solids **149**, 381 (1999).
37. V. Dierolf, H. Paus, and F. Luty, Phys. Rev. B **43**, 9879 (1991).
38. M. Krantz, F. Luty, V. Dierolf, and H. Paus, Phys. Rev. B **43**, 9888 (1991).
39. H. Söthe, J.-M. Spaeth, and F. Luty, Rev. Solid State Sci. **4**, 440 (1990).
40. H. Söthe, J.-M. Spaeth, and F. Luty, Radiat. Eff. Defects Solids **119–121**, 269 (1991).
41. H. Söthe, J.-M. Spaeth, and F. Luty, J. Phys.: Condens. Matter **5**, 1957 (1993).
42. J.-M. Spaeth, J. Niklas, and R. Bartram, *Structural Analysis of Point Defects in Solids*, Springer Series in Solid-State Sciences Vol. 43 (Springer, Berlin, Heidelberg, 1992).
43. V. Dierolf and J.-M. Spaeth, Mat. Science Forum (Proc. ICDIM96) **239–241**, 461 (1997).
44. T. Pawlik, R. Bungenstock, J.-M. Spaeth, and F. Luty, Radiat. Eff. Defects Solids **134**, 465 (1195).
45. F. M. M. Yasuoka, J. C. Castro, and L. A. O. Nunes, Phys. Rev. B **43**, 9295 (1991).
46. C. Sandmann, Master's thesis, Universität Paderborn, 1999.
47. M. Koerdt, Master's thesis, Universität Paderborn, 1998.
48. A. Ostendorf, Master's thesis, Universität Paderborn, 2000.
49. V. Dierolf and M. Koerdt, Phys. Rev. B **61**, 8043 (2000).
50. V. Dierolf, A. Kutsenko, C. Sandmann, G. Corradi, and T. Tröster, J. Lumin. **87–89**, 989 (2000).
51. V. Dierolf, Phys. Rev. B **60**, 12601 (1999).
52. D.-J. Jang and J. Lee, Solid State Commun. **94**, 539 (1995).
53. E. Gustin, M. Leblans, A. Bouwen, and D. Schoemaker, Phys. Rev. B **54**, 6977 (1996).

2 F Centers and Rare-Earth Ion Defects in Alkali Halides

Before we consider the effects of interaction between electronic defect states and molecular defects, we give in this chapter a brief overview of the basic properties of the defects considered in the remaining chapters of this book. As in the rest of the text we concentrate on F centers and divalent rare-earth ions. The properties of these defects have been reported in several other monographs, and we shall review here only those aspects that are important in the context of their interaction with molecular defects.

2.1 F-Center-Related Defects

The F center in alkali halides is still the best studied [1] and understood prototype and model case as a simple one-electron defect and was therefore the starting point for investigations of electronic–molecular defect centers. The F center is characterized by a very strong electron–phonon coupling which results in two drastically different electron–lattice states (see Fig. 2.1): a deep, compact 1s-type ground state(GS) and a rather extended 2s/2p mixed relaxed excited state (RES) lying very close ($\sim$ 0.1 eV [2]) to the conduction band (CB). The physical properties of this unstable, shallow RES make photoexcited F centers very versatile in terms of the many possible subsequent processes: while at the lowest temperatures T_0 optical excitation leads to

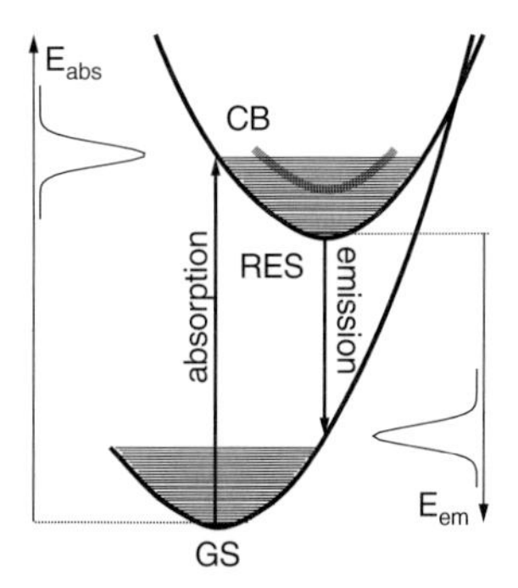

Fig. 2.1. Schematic configuration coordinate diagram for an F center. The absorption and emission spectra, with their expected Gaussian line shape, are indicated, with energy scales starting at the ground state (GS) and relaxed excited state (RES) level respectively. For the case where the electron is in the RES the conduction band (CB) is drawn schematically in gray assuming that its dependence on the configuration coordinate is the same as for the RES

a Stokes-shifted strong luminescence with a long lifetime of the order of microseconds, in a slightly higher temperature range ($T_1 = 80$–160 K, depending on the host), the electron can be excited thermally from the RES into the conduction band leaving behind an empty anion vacancy F^{ion}. The electron can in turn be trapped by another F center forming an F′ center.

At still higher temperatures ($T_2 > 170$ K, in Cs halides, and $T_2 > 220$ K in other alkali halides), the F^{ion} can undergo vacancy diffusion and become trapped either by another F center or at a cationic or anionic point defect which has been intentionally doped into the crystal. Using these processes of diffusion and trapping, a large variety of F aggregate centers can be formed by light irradiation in the F absorption band at a temperature T_2. This procedure is often referred to as "aggregation". The resulting center complexes have been investigated in terms of their interesting physical properties as well as their potential for applications (e.g. optical information storage [3] and tunable lasers [4]). In combination with OH^- and CN^- molecular defects, the flexibility of the F center and its electron makes it possible to create

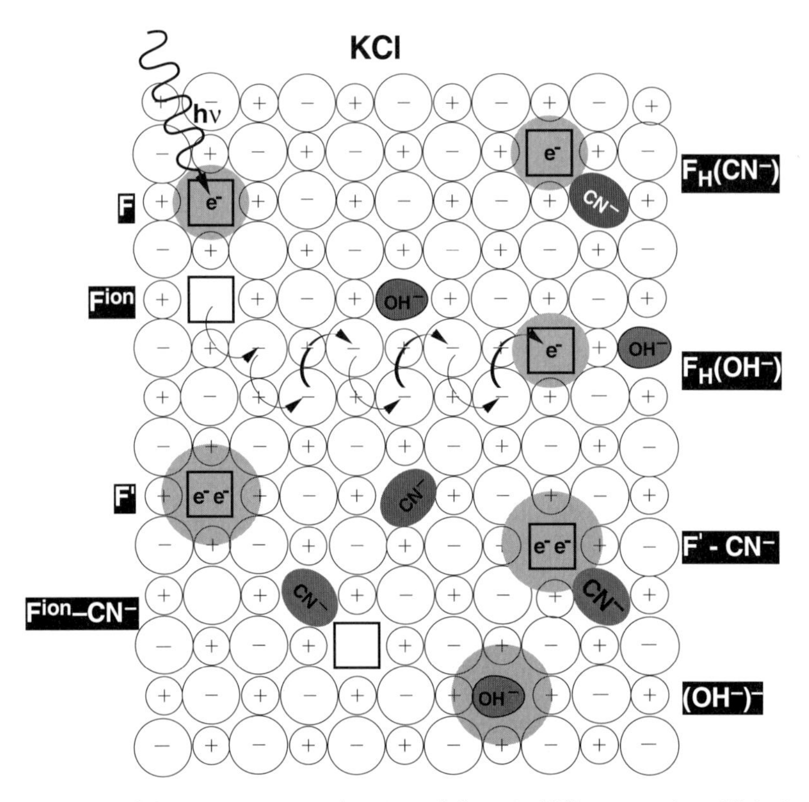

Fig. 2.2. Schematic picture of various defects in KCl present in additively colored samples. The diffusion of anion vacancies and trapping at a defect site are indicated by *curved arrows*

a great variety of electronic–molecular defects in a very controlled way and investigate them individually. These electronic–molecular defects include:

- F centers with a molecule as their ⟨100⟩, ⟨110⟩, or ⟨200⟩ neighbor, i.e. $F_H(CN^-)$ and $F_H(OH^-)$ centers.
- F centers with two molecules as neighbors, e.g. $F_{H2}(OH^-)$.
- Molecules located next to an anion vacancy occupied by no electrons, (e.g. F^{ion}–CN^-), or two electrons, (e.g. F'–CN^-).
- Molecular defects which have captured an extra electron, e.g. $(OH^-)^-$

Some of these defects are shown schematically in Fig. 2.2 for the fcc (NaCl type) host materials.

2.2 Divalent Rare-Earth Ions in Alkali Halides

The rare-earth (RE) ions play an important role as luminescing ions in insulating materials and are often used as active ions in solid-state lasers. Their most common charge state is the trivalent positive one which is unfavorable for incorporation into alkali halides. However, Eu^{2+}, Yb^{2+}, and Sm^{2+} also exist as divalent ions with high enough ionization energies to be stable. Individual RE^{2+} defects in various crystalline materials have been studied for a long time, and their spectroscopic properties were first summarized by McClure and Kiss [5]. A more recent survey, which concentrates on halide crystals and also includes results from magnetic-resonance studies has been published by Rubio [6]. It gives an excellent list of further reading.

2.2.1 Isolated Yb^{2+} Ions

Background. Figure 2.3 summarizes the properties of the Yb^{2+} electronic states and the corresponding optical transitions which are most important in our context. It is well established for the free Yb^{2+} ion [7] that the lowest allowed optical transitions from its closed-shell $4f^{14}$ ground state take place (with different relative strengths) into the numerous sublevels of the $4f^{13}d^1$ excited state. When substituted as a defect in an ionic crystal, the strong influences of the crystal surrounding through electrostatic forces, exchange interaction, overlap, and covalency must be considered. These contributions are summarized by the "crystal field" and can be expressed in the case of cubic symmetry by a single parameter D_q. In alkali halide hosts, the substitutional O_h site symmetry of Yb^{2+} is, in principle, lowered to C_{2v} owing to the charge-compensating cation-vacancy on a neighboring site in the ⟨110⟩ direction. However, the perturbation on the optical transition caused by this vacancy is small, because the electronic orbitals of the excited state interact predominantly with the six oppositely charged nearest-neighbor anions

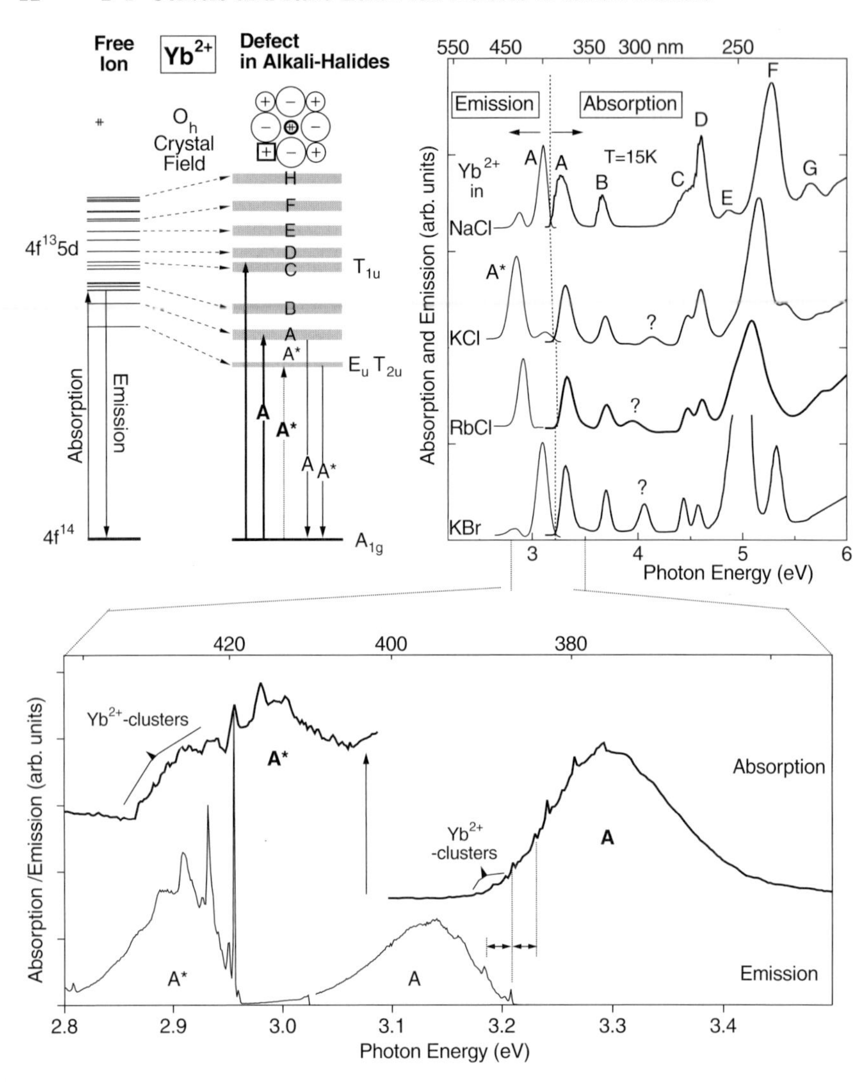

Fig. 2.3. $4f^{14} \leftrightarrow 4f^{13}d^1$ electronic transitions of Yb^{2+}: (*left*) schematic energy-level diagram of free and isolated Yb^{2+}ions in alkali halides; (*right*) absorption and emission spectra in various hosts; (*bottom*) detailed absorption and emission spectra in KCl

and can be (and usually is) neglected to first order. Consequently, theoretical treatments with an O_h crystal field are quite appropriate [8, 9], and yield very similar results to those for Yb^{2+} without a vacancy in a divalent ionic crystal [10], such as $SrCl_2$.

Spectroscopic Properties. Figure 2.3 (left) illustrates schematically the results of calculations [8, 10, 11] with an O_h crystal field of an appropriately chosen strength. According to Hund's rule the $4f^{14}$ ground state is of A_{1g} symmetry and therefore only transitions to the 18 (5d) states with T_{1u} symmetry are electric-dipole allowed and give contributions to the absorption spectrum. We indicate (following [9]) these phonon-broadened bands by bold letters **A**,**B**,**C**,... in the case of absorption and regular letters in the case of emission. The lowest energy $^3P_2$ term forms two E_u and T_{2u} states with a very small separation and hence absorption (**A***) and emission (A*) transitions between A_{1g} and these states should, in principle, be forbidden.

In Fig. 2.3 (right), the resulting optical transitions in the UV/visible absorption spectra of Yb^{2+} defects for low Yb^{2+} concentrations in four different hosts at 15 K are depicted [12, 8, 9, 13]. A characteristic feature of all the spectra is a multiband structure (**A,B,C,...**) which is expected from the above considerations. When the host material is varied (NaCl, KCl, RbCl, and KBr) the spectra differ only by small shifts or broadening effects. This finding is reflected in the O_h crystal field strength parameter D_q, which varies only slightly (around $D_q \approx 1200$ cm^{-1}) in these four hosts [8]. Exceptions from the good agreement with the predictions are the observed band **G** and the band between **B** and **C** indicated by a question mark, which cannot be fitted by this simple crystal field model. Their different origin is confirmed by the observation (Fig. 2.3) that their position and strength change greated when the host is varied. The behavior of these bands is not clear yet, but it is speculated [8] that they are related to covalency effects neglected in the crystal field analysis. The importance of these contributions to the crystal field will become more apparent when the drastic changes in optical transitions caused by neighboring CN^- ions are presented and discussed in Sect. 9.2.

The emission properties of Yb^{2+} in NaCl have been observed and interpreted [9, 14] by Tsubui et al. At low temperatures, the spectrum consists of two emission bands A and A* (called bands I and II in [14]). The strong emission band A is the mirror image of the absorption band **A** around a common zero-phonon line. Separated from the A emission by $\sim$ 2000 cm $^{-1}$ or 250 meV there appears a weak emission band (A*). The qualitative interpretation in the case of NaCl can be easily understood from the level scheme in Fig. 2.3 (right), as a "three-step process" occurring after optical excitation in any of the higher absorption bands (**B**, **C**, $\cdots$):

(i) The system relaxes into the lowest T_{1u} state by multi-phonon relaxation processes [15].
(ii) In this state, the A emission rate competes with nonradiative relaxation into the E_u or T_{2u} state.
(iii) The small fraction of ions which have relaxed into the E_u and T_{2u} states exhibit a weak A* emission of long radiative lifetime (of the order of milliseconds).

An extension of these emission studies [12] to other host materials showed basically the same behavior in KBr as in NaCl with an A emission much stronger than A* (Fig. 2.3, right). In contrast to this, the A $\rightarrow$ A* relaxation is obviously much more efficient in KCl and RbCl, eliminating nearly totally the A emission at the cost of a strong A* emission. The observation of this symmetry-forbidden A* emission (E_u and $T_{2u} \rightarrow A_{1g}$) suggests that the opposite **A*** absorption process should be observable for Yb^{2+} in NaCl, KCl, and KBr hosts as well, and indeed, the process could be observed using much thicker crystals with higher Yb^{2+} concentrations ($\approx 10^{-3}$).

We shall illustrate this for KCl, the host which will be used most often as an example in the following. In order to identify and correlate corresponding emission and absorption peaks, the spectral range of the A and A* transition is shown in Fig. 2.3 (bottom) for this host on an extended scale. Unlike the low-resolution survey measurements shown in the top of the figure, the higher spectral resolution, here reveals a resolved vibronic substructure both in absorption and in emission; this has been measured and analyzed for the **A** band in terms of coupling to pseudo-localized modes [13]. For both the **A*** absorption and the A* emission, a similar vibronic structure is found reflecting a coupling to modes with frequencies which are equal within experimental error ($\sim$ 210 cm^{-1} in KCl). The electron–phonon coupling constant can be estimated from comparison of the ratio of strengths of the zero-phonon emission and the total emission I_{tot} using the following relation [16]:

$$\frac{\int_{\text{ZPL}} I(E)\mathrm{d}E}{\int_{\text{tot}} I(E)\mathrm{d}E} = \mathrm{e}^{-S(T)} .$$

For the A and A* emissions in KCl we find Huang–Rhys factors of $S_A \approx 6$ and $S_{A*} \approx 3$, respectively, which are much lower than the ones encountered for F centers. Close examination of the absorption and emission spectra, which should be mirror images of each other, reveals that in the **A** and **A*** absorption region, in addition to the isolated Yb^{2+} ions, further Yb^{2+}-related defects are present. While barely visible on the long-wavelength side of the **A** absorption, a clear band with substructure appears in the **A*** absorption, indicating that the integrated absorption strength ratio A*/A of this additional center is drastically different from that of the regular Yb^{2+} defect. The extra bands are not present in the emission spectra, which were obtained for well-quenched samples with a low Yb^{2+} concentration, suggesting that these extra absorption bands are related to Yb^{2+} pairs or more complex Yb^{2+} aggregates. The strong dependence of the A*/A ratio on the type of defect center which is in contrast to the weak influence on the spectral positions, is further supported by the strong variation found for the different host materials (NaCl, 500; KCl, 3000; and KBr, 1000). This sensitivity of the relative oscillator strengths to the *static* surroundings becomes even more pronounced when *dynamic* interaction effects are considered in Sect. 9.9.

2.2.2 Isolated Eu^{2+} Ions

Background. Among the divalent rare earth defects the Eu^{2+} ion has been studied must intensively in the past in numerous ionic crystals including the alkali halides [6]. Due to the high quantum efficiency of its blue emission it has been considered as a possible ion for laser operation in the spectral region. Although this expectation has not been fulfilled so far, many other applications are based on the luminescing properties of this ion. For instance, X-ray storage phosphors have been developed and commercialized in the BaFBr host material. Also in alkali halides, applications e.g.: as optical memory [17] and dosimeters [18, 19] have been proposed and demonstrated.

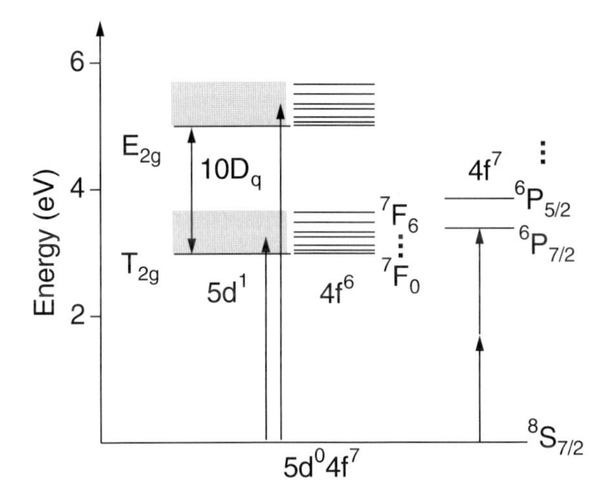

Fig. 2.4. Simplified level scheme of Eu^{2+}in alkali halides. The one-photon $4f \rightarrow 5d$ and two-photon $4f \rightarrow 4f$ absorption transitions are shown by *arrows*

Spectroscopic Properties. The optical properties of Eu^{2+} ion in ionic crystals have been studied in detail over the last two decades [20, 21, 22, 23, 24, 25, 26] so that the nature of the transitions involved is well understood. The electronic configuration $4f^7 4s^2 6p^6$ with its half-filled 4f shell results in a $4f^7$ ($^8S_{7/2}$) ground state and two $5d^1 4f^6$ excited levels labeled T_{2g} and E_g according to their symmetry in a cubic crystal field. Owing to the interaction with the $4f^6$ orbitals, spin–orbit interaction, electron–electron interaction, and the lower symmetry of the crystal field, the 5d levels are split. The influence of the different contributions has been a subject of controversy. However, in many host materials, seven subbands can be identified, suggesting that the 5d–4f interaction plays a major role. The simplified case of a very weak interaction, for which the combined $5d^1$ $4f^6$ states are essentially just a linear combination of the independent $5d^1$ and $4f^6(^7F_0 ... ^7F_6)$ states is depicted in Fig. 2.4. This energy level scheme is reflected in the absorption spectrum (shown by a dotted line for KCl in Fig. 10.1) by two main bands located in the near UV with a "staircase" type substructure this most noticeable in low-energy transitions. Owing to the electric-dipole-allowed character, the interconfigurational transitions between the $4f^7$ ground state and the $5d^1\ 4f^6$ excited state are dominant in the absorption spectra, intraconfigurational transitions within the $5d^1 4f^6$ shell are possible as well. This has been shown most convincingly by Nunes et al. [25] by two-photon absorption experiments. The energy levels involved are shown in Fig. 2.4.

2.2.3 Isolated Sm^{2+} Ions

Background. The excitation–relaxation cycle of an isolated Sm^{2+} defect in KCl, consisting of an Sm^{2+} ion and a charge-compensating vacancy as its ⟨110⟩ neighbor [27, 28], is illustrated in Fig. 2.5, using schematic energy-level diagrams[1] proposed by Guzzi and Baldini [16]. After excitation into the 5d-type excited state and rapid relaxation into the (4f-type) $^5D_0$ state, transitions into the ^7F_j states take place giving rise to an emission spectrum which at low temperatures ($T < 15$ K) consists of seven groups of sharp lines stretching from the visible into the near infrared. Owing to the $4f \leftrightarrow 4f$ character of the transitions, the lifetime is quite long ($\approx$ 10 ms). At higher temperatures, a broad emission background with a much higher decay rate

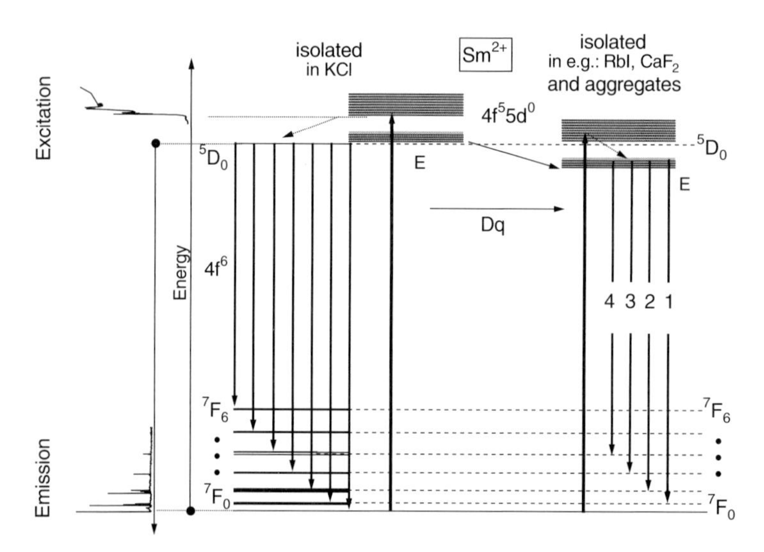

Fig. 2.5. Schematic energy level diagram for the lowest states of isolated Sm^{2+} in KCl, and for isolated Sm^{2+} in some other hosts such as RbI and CaF_2 and for Sm^{2+} aggregates in KCl. On the *left*, the corresponding excitation and emission spectra are shown

[1] An alternative, more detailed description proposed by Fong [29] will be used later in Chap. 11.

appears indicating a thermally activated population of a 5d-type state (E), which lies only slightly (0.016 eV for KCl) above the 4f-type states. Owing to its 5d character, this level is fairly sensitive to changes in the host material and for RbI, KI, NaCl, NaBr, NaI, and CaF_2 [6, 16] it even lies below the $^5D_0$ state (Fig. 2.5, right). For these materials the emission properties are altered: the luminescence is spectrally much broader and has, even at low temperature, a much shorter radiative lifetime and a high quantum efficiency. The latter features have been used successfully in the CaF_2 host material for the realization of a solid-state laser [30].

The electron–phonon interaction of the Sm^{2+} ion in KCl was studied experimentally by Baldini and Guzzi [31] and by Bron and Heller [28], who found a significant difference in the strength and spectral shape of the vibrational sidebands. These observations were explained by the difference in the nature of the coupling [32]: while the 4f $\leftrightarrow$ 4f transitions exert only long-range forces, which cause exclusively an interaction with unperturbed crystal phonons (although this interaction is weak and corresponds to $S \approx 1$), the 4f $\leftrightarrow$ 5d transitions involve also short-range contributions to the force, which cause an interaction with local (impurity-perturbed) phonon modes ($S \approx 5$).

Spectroscopic Properties. The absorption, excitation and emission spectra of Sm^{2+} in KCl have already been studied in great detail [27, 33] thus providing an excellent basis for the present study. Of particular importance for investigations of the effects of interaction with molecules described here was the ability to identify the various Sm^{2+} and $(Sm^{2+})_n$-related defects (with weak emission responses) which are always present, and to distinguish them from the Sm^{2+}–$(CN^-)_n$ defects, which were the primary object of the investigation. Combined excitation–emission spectroscopy is an excellent tool for performing this task. In this technique, applications of which will be described on various occasions in the course of this book, a large number of emission spectra are recorded for a dense sequence of excitation energies. The resulting 2D dataset of emission intensities as a function of the emission and excitation energies is best visualized as a contour plot.

As a good example of the usefulness and sensitivity of the CEES technique and for comparison later on, Fig. 2.6 shows[2] a contour plot obtained from emission spectra for a KCl sample doped only with Sm^{2+}. The spectral regions shown were selected to include just a *single* peak for each type of center, i.e. the individual $^7F_0 \rightarrow {}^5D_0$ transition in excitation and the line with the highest energy out of the crystal-field-split $^5D_0 \rightarrow {}^7F_0$ transitions in emission. Hence, each peak ("mountain") found in the contour plot of the CEES measurement corresponds to a *different* center type. The strongest peaks (*a* and *d*) are labeled as in [27] where site *a* has been identified as the main site, which consists of a Sm^{2+} ion with a charge-compensating cation

[2] The corresponding overall appearance of the emission and excitation spectra is essentially identical to that for CN^--codoped samples shown in Fig. 11.1.

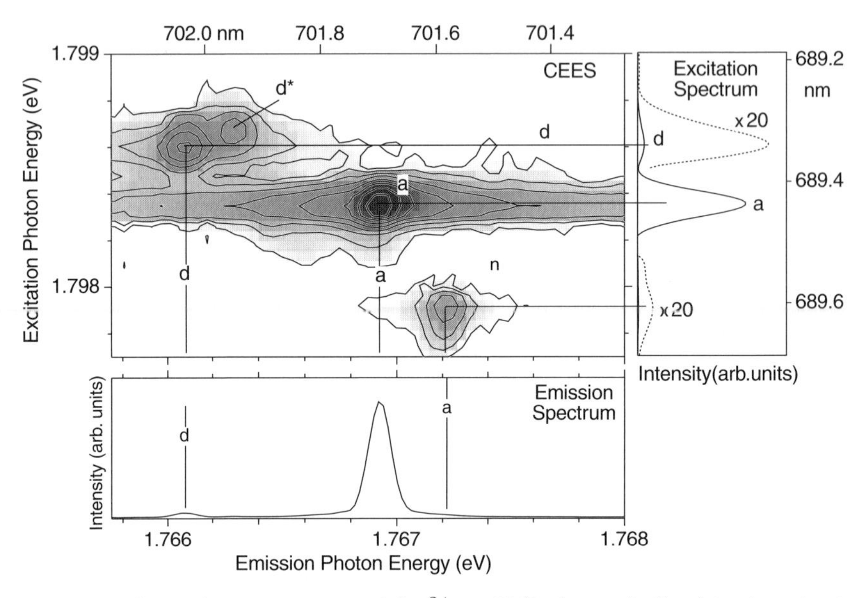

Fig. 2.6. Site-selective spectra of Sm^{2+} in KCl: (*center*) Combined excitation–emission spectrum, (*bottom*) emission spectrum for a broad band excitation around 689 nm, (*left*) excitation spectrum of emission probed with low spectral resolution (702 nm ± 2 nm). The various sites are labeled as in [27]

(K) vacancy as one of its next-nearest neighbors in the ⟨110⟩ direction. These two sites a and d can also be observed easily in individual emission and excitation spectra as shown in Fig. 2.6 (bottom and right). A closer look at the combined spectrum reveals at least two more peaks. These peaks are about two orders of magnitude weaker than the main peak. While one of them (d^*) can easily be obscured by the emission from site d, the other one (n) is well separated but is so weak that it was not among the peaks reported (a–m) in [27].

(Sm^{2+})$_n$-Complexes. Before we turn our attention to the CN^- related Sm defect complexes (see Chap. 11) we consider first Sm^{2+} complexes that consist of several Sm^{2+} ions. These complexes provide a good example of how the ion reacts to changes in its environment. Similarly to other divalent cationic defects in alkali halides, the Sm^{2+} ion has a tendency to form aggregates when the crystal is stored for some time at room temperature. Even after careful quenching procedures, a small number of these $(Sm^{2+})_n$ complexes remain [27]. The importance of the quenching procedure is reflected by the emission and excitation spectra of a KCl:Sm^{2+} sample which has been stored several months in the dark at room temperature. Under these conditions a center aggregation can take place [34] which reveals itself by a drastically changed emission spectrum (Fig. 2.7 left). This change in the emission spectrum can be observed for a wide range of excitation energies (Fig. 2.7 right)

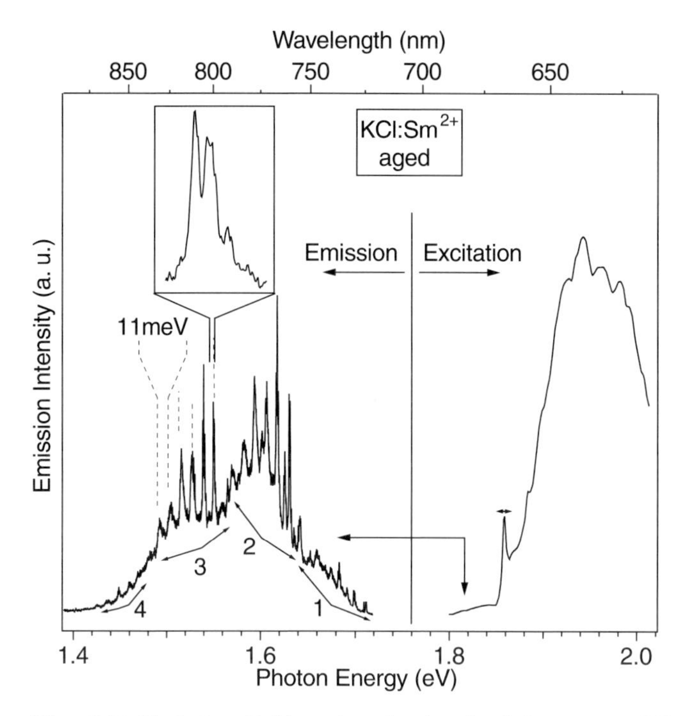

Fig. 2.7. Emission (*left*) and excitation (*right*) spectra of Sm^{2+} in a KCl sample which was stored for several weeks at room temperature. The emission was excited at the spectral position indicated by the arrow at 1.8 eV. *Inset*: substructure of the indicated emission line. The excitation spectrum was probed in the spectral range of the inset ($\approx$ 800 nm)

and disappears after the sample is quenched. The emission spectrum consists of a broad background, on top of which several groups (1–4) of rather sharp lines are situated. Each of these groups is, in turn, composed of several lines separated by $\approx$ 11 meV. In the excitation spectrum of the emission, probed at around 1.55 eV or 800 nm, one finds a broad band with a substructure between 1.90 eV (650 nm) and 1.85 eV (670 nm) consisting of one sharp and few less pronounced maxim. This suggests an electron–phonon coupling similar to that observed for the transitions to 5d-type states in isolated Sm^{2+}, but shifted to considerably lower energies. The emission spectrum strongly resembles those in NaI and NaBr shown in [16]. For that reason, we conclude that in these aggregated center types the level E is also shifted below the $^5D_0$, state as schematically shown in Fig. 2.5.

This assumption is supported by the red-shifted 5d-type excitation spectrum. In this interpretation, the observed groups (1–4) of emission bands indicated in Fig. 2.7 are assigned to the E $\rightarrow$ ^7F_j (j = 1...4) transitions. The sharp, equidistant peaks on top of the bands can be identified as a vibrational substructure caused by coupling to a localized mode with a frequency

($\overline{\omega}_{\text{loc}} \approx 11$ meV) considerably lower than that of the mode (≈ 26 meV) found for isolated Sm^{2+} defects [13]. This indicates that the surroundings of the Sm^{2+} are changed strongly within these complexes compared with the isolated Sm^{2+}.

On further spectral expansion of the emission spectrum (see inset) an additional substructure becomes apparent, a clear indication that several distinctively different types of center exist.

While these defect centers in samples doped only with Sm^{2+} are very interesting in themselves and the Sm^{2+} aggregates have not been studied in complete detail so far, in this book we use the spectra of isolated Sm^{2+} and aggregated Sm^{2+} complexes that have been published, and those presented above, merely for reference and for comparison with the results discussed in Chap. 11 for samples codoped with CN^-. This allows a reliable identification of the Sm^{2+}:$(CN^-)_n$-related centers.

References

1. *Physics of Color Centers*, edited by W. B. Fowler (Academic Press, New York, 1968).
2. Y. Kondo and H. Kanzaki, Phys. Rev. Lett. **20**, 790 (1975).
3. H. Blume, T. Bader, and F. Luty, Opt. Commun. **12**, 147 (1974).
4. W. Gellermann, J. Phys. Chem. Solids **52**, 249 (1991).
5. D. S. McClure and Z. Kiss, J. Chem. Phys. **39**, 3251 (1963).
6. J. O. Rubio, J. Phys. Chem. Solids **52**, 101 (1991).
7. W. Bryant, J. Opt. Soc. Am. **55**, 771 (1965).
8. S. W. Bland and M. J. A. Smith, J. Phys. C: Solid State Phys. **18**, 1525 (1985).
9. T. Tsubui, H. Witzke, and D. S. McClure, J. Lumin. **24&25**, 305 (1981).
10. T. S. Piper, J. P. Brown, and D. S. McClure, J. Chem. Phys. **46**, 1353 (1967).
11. M. V. Eremin, Opt. Spectrosc. **29**, 53 (1970).
12. C. P. An, V. Dierolf, and F. Luty, Phys. Rev. B **61**, 6565 (2000).
13. M. Wagner and W. E. Bron, Phys. Rev. **139**, 223 (1965).
14. T. Tsubui, D. S. McClure, and W. C. Wong, Phys. Rev. B **48**, 62 (1993).
15. A. Riseberg and H. W. Moos, Phys. Rev. **114**, 429 (1968).
16. M. Guzzi and G. Baldini, J. Lumin. **6**, 270 (1973).
17. H. Nanto, K. Murayama, T. Usuda, F. Endo, Y. Hirai, S. Taniguchi, and N. Takeuchi, J. Appl. Phys. **74**, 1445 (1993).
18. R. Melendrez, R. Perez-Salas, L. P. Pashchenko, R. Aceves, T. M. Piters, and M. Barboza-Flores, Appl. Phys. Lett. **68**, 3398 (1996).
19. I. A. de Carcer, F. Cussó, F. Jaque, E. Espana, T. Calderon, G. Lifante, and P. D. Townsend, J. Phys. D: Appl. Phys. **26**, 154 (1993).
20. R. Reisfeld and A. Glasner, J. Opt. Soc. Am. **54**, 331 (1964).
21. J. Kirs and A. Niilish, Trans. Inst. Fiz. Astron. Akad. Nauk. Est. SSR **18**, 36 (1962).
22. J. A. Hernandez, W. K. Cory, and J. O. Rubio, J. Chem. Phys. **72**, 198 (1980).
23. J. A. Hernandez, F. J. Lopez, H. S. Murrieta, and J. O. Rubio, J. Phys. Soc. Japan **50**, 225 (1981).

24. W. E. Bron and M. Wagner, Phys. Rev. **139**, 233 (1965).
25. L. A. O. Nunes, F. M. Matinaga, and J. C. Castro, Phys. Rev. B **32**, 8356 (1985).
26. D. Cassi, M. Manfredi, and M. Solzi, Phys. Status. Solidi B **135**, K143 (1986).
27. A. J. Ramponi and J. C. Wright, Phys. Rev. B **31**, 3965 (1985).
28. W. E. Bron and W. R. Heller, Phys. Rev. **136**, 1433 (1964).
29. F. K. Fong, *Theory of Molecular Relaxation: Applications in Chemistry and Biology* (Wiley, New York, 1975).
30. W. Kaiser, C. G. W. Garrett, and D. L. Wood, Phys. Rev. **123**, 766 (1961).
31. G. Baldini and M. Guzzi, Phys. Status Solidi **30**, 601 (1968).
32. E. Mulazzi, G. F. Nardelli, and N. Terzi, Phys. Rev. **172**, 847 (1968).
33. V. Dierolf, Phys. Rev. B **60**, 12601 (1999).
34. A. J. Ramponi and J. C. Wright, Phys. Rev. B **35**, 2413 (1987).

3 Properties of Molecular Defects

Molecular defects are commonly described as anharmonic oscillators, the basic properties of which have been well studied for a long time [1] and most aspects have been treated in textbooks [2]. Still, it is useful to review some of these points briefly and to discuss their applicability to the defect systems under consideration here.

3.1 Transition Energies

3.1.1 Free Molecules

The parabolic potential of a simple harmonic diatomic oscillator can be expressed as

$$U_{\mathrm{harm}} = \frac{1}{2}\mu \left(2\pi c\overline{\omega}_{\mathrm{e}}\right)^2 r_{\mathrm{e}}^2 \xi^2 \,, \tag{3.1}$$

where $\xi = \frac{r - r_{\mathrm{e}}}{r_{\mathrm{e}}}$ is the normalized relative distance between the atoms which have an average separation r_{e}, masses m_1, m_2, and a reduced mass $\mu = \frac{m_1 m_2}{m_1 + m_2}$. The corresponding energy eigenvalues of the Schrödinger equation have the well-known form

$$\frac{E_v}{hc} = \overline{\omega}_{\mathrm{e}} \left(v + \frac{1}{2}\right) , \tag{3.2}$$

where $\overline{\omega}_{\mathrm{e}}$ is the harmonic eigenfrequency in wave numbers[1]. Despite being a good first approximation, the above expressions need to be refined by including higher order terms (ξ^3, ξ^4, ...) in a Taylor series for the potential to give it a more realistic shape. These additional terms are collectively referred

[1] Several (energy) units are used to describe transition energies in molecular spectroscopy: wavelength (μm), photon energy E_v (eV), and frequency $\overline{\omega}_{\mathrm{e}}$ (cm^{-1}). The latter is especially confusing because the name and the unit are not consistent. For this reason, a bar is used to distinguish $\overline{\omega}$ (cm^{-1}) from the regular frequencies ω, which have the unit s^{-1}. These quantities are related to each other by $\hbar\omega = hc\overline{\omega}$. For readers' convenience and to allow easy comparison with result for the VIS, several commonly used scales for the transition energy are used simultaneously in most of the IR spectra presented here.

to as "mechanical anharmonicity", and under the assumption that they are small, the energy levels of a (mechanical) anharmonic oscillator can be written as

$$\frac{E_v}{hc} = \overline{\omega}_e \left(v + \frac{1}{2}\right) - \overline{\omega}_e x_e \left(v + \frac{1}{2}\right)^2 + \overline{\omega}_e y_e \left(v + \frac{1}{2}\right)^3 + \dots . \tag{3.3}$$

The Taylor series of the potential usually converges only after many terms yielding a large number of adjustable parameters and a complicated relation to $\overline{\omega}_e$ and $\overline{\omega}_e x_e$, because no physical insight about the expected shape has been used. This disadvantage is overcome by use of simplified empirical potentials among which the Morse potential, [3] with only two parameters U_D (dissociation energy) and β has been found to be a rather good description for most weakly anharmonic oscillators. This potential can be expressed in the following form:

$$U(\xi) = U_D \left[1 - \exp(-\beta\xi)\right]^2 . \tag{3.4}$$

For this potential, the Schrödinger equation can be solved and it turns out that the cubic and higher terms in (3.3) are exactly zero [3], allowing an easy calculation of the transition energies. The parameters U_D, β on one hand and $\overline{\omega}_e$, $\overline{\omega}_e x_e$ on the other are related as follows:

$$\begin{aligned} \overline{\omega}_e &= \beta \cdot \sqrt{\frac{\hbar \cdot U_D}{\pi \cdot c \cdot \mu}} \propto \frac{1}{\sqrt{\mu}} , \\ \overline{\omega}_e x_e &= \frac{\hbar \cdot \beta^2}{4 \cdot \pi \cdot \mu} , \\ x_e &= \frac{\hbar \cdot \omega_e}{U_D} \propto \frac{1}{\sqrt{\mu}} . \end{aligned} \tag{3.5}$$

It follows from (3.5) that both the harmonic eigenfrequency $\overline{\omega}_e$ and the anharmonic parameter x_e depend in the same way on the reduced mass and therefore both should change under isotope substitution (as long as the Morse potential does not change) by the square root of their reduced mass ratio. This isotope effect is most pronounced for OH^- and OD^-, where $\sqrt{\frac{\mu_{OD}}{\mu_{OH}}} = 1.375$.

3.1.2 Transition Energies of Molecular Ions in a Solid

When a molecule is incorporated into a crystal, the molecular potentials are changed, leading to a shift in eigenfrequency. As long as the shift is small,[2] the influence of the host lattice ions on the potential can be treated in perturbation theory as an additional small potential which can be divided into two parts:

[2] The most pronounced exception to this assumption is the OH^- molecule, which shares an anion vacancy with an extra electron, and will be treated in Sect. 8.2.

(i) An attractive potential, which is produced mostly by Coulomb and electric multipole interactions between the molecular ion and the host ions. In ionic crystals it has the tendency to pull the atoms of the molecule apart and therefore leads to a shift of the eigenfrequency to lower energy (a "redshift").
(ii) A repulsive potential, produced by an overlap of the electronic distributions of the molecule and of the host ions. This interaction pushes the atoms of the molecule together resulting in an eigenfrequency shifted to higher energy (a "blue" shift).

For both of the isolated[3] substitutional molecular defects in alkali halides, namely OH^- and CN^-, considered here, it has been found that the blueshift of the vibrational frequency increases with decreasing size of the lattice ion and decreasing lattice constant [4, 5, 6, 7], showing that the repulsive term dominates. Using a Born, Mayer and Huggins-type potential [8, 9, 10] and neglecting the attractive contributions, it can be shown that the fundamental transition frequency $\overline{\omega}_{1\leftrightarrow 0}$ of a molecule embedded in an alkali halide crystal can be expressed as

$$\overline{\omega}_{1\leftrightarrow 0} = \overline{\omega}^{\text{free}}_{1\leftrightarrow 0} + \Omega \exp\left(-\frac{d}{\rho}\right) , \tag{3.6}$$

where Ω and ρ are molecule-dependent parameters and d is the interionic distance in the host. By fitting the experimental data of 16 alkali halides for CN^-, An and Luty [7] found very good agreement with (3.6) using the parameters $\overline{\omega}^{\text{free}}_{1\leftrightarrow 0}$ and Ω, ρ listed in Table 3.1 and d, the lattice constant of the host material. [7] For OH^-/OD^-, a good fit could be obtained only if d was replaced by $\sqrt{2}a$ for a host with the NaCl crystal structure and by $\sqrt{3}a$ for the CsCl structure. For the anion radii a, the values from Bosi and Nimis [11] were used.

Table 3.1. Fitting parameters for (3.6) for CN^-, OH^-, and OD^- molecules in alkali halides

Molecule	$\overline{\omega}^{\text{free}}_{1\leftrightarrow 0}$ (cm^{-1})	Ω (cm^{-1})	ρ (Å)
CN^-	2027	742	1.23
OH^-	3546.7	1375.2	0.987
OD^-	2619.9	886.5	1.005

More empirically an Ivey-type law has been proposed,

$$\overline{\omega}_{1\leftrightarrow 0} = \overline{\omega}^{\text{free}}_{1\leftrightarrow 0} + kV^{-2/3} \tag{3.7}$$

[3] Within this book "isolated" refers to the case in which the molecule is embedded in the lattice as a substitutional defect but is not further disturbed by other defects. In contrast, we use "free" for a molecule in vacuum (i.e. outside the crystal).

which relates the energy shift to the amount of "available volume" (V) for the molecule [12]. In both relations, the eigenfrequency $\overline{\omega}^{\rm free}_{1\leftrightarrow 0}$ of the free molecule yields the limiting lowest frequency. While this value has not been measured directly for CN^- ions, it has been determined for OH^- by means of photodetachment [13] and found to be given by, $\overline{\omega}^{\rm free}_{1\leftrightarrow 0}(OH^-) = 3555\ {\rm cm}^{-1}$, fairly close to the value ($3546.7\ {\rm cm}^{-1}$) extrapolated from the data using (3.6).

As we shall see, the spectral changes that are observed for the defect system under consideration in this work can often be explained on this basis. However, two remarkable deviations are found for Yb^{2+}–$(CN^-)_n$ and $(OH^-)^-$, which are treated in Sects. 9.3 and 8.2, respectively.

3.2 Absorption Intensity and Oscillator Strengths

3.2.1 Free Molecule

Considering first the harmonic oscillator again, we can write the integrated absorption strength Q_{01} as

$$Q_{01} = \frac{Ne^2\pi}{c^2\mu} f \,, \tag{3.8}$$

where we have introduced the dimensionless oscillator strength as

$$f = \frac{1}{3}\left(\frac{1}{e}\frac{{\rm d}p}{{\rm d}\xi}\right)^2 . \tag{3.9}$$

Here μ is the reduced mass, c the speed of light, N the number of oscillators per unit volume, e the electronic charge and $\frac{{\rm d}p}{{\rm d}\xi}$ the first derivative of the dipole moment function (all in CGS esu units). For the molecules under consideration (CN^-, OH^-, and OD^-) the oscillator strength are included in Table 3.2.

For a harmonic oscillator which has a linear dipole moment function, the $0 \to 1$ fundamental transition is the only one which is expected to be observable in IR absorption at sufficiently low temperatures. This changes as soon as anharmonicity is taken into account. *Two* contributions have to be considered:

- *mechanical anharmonicity* due to a nonparabolic potential, and
- *electric anharmonicity* due to a nonlinear dependence of the dipole moment on the normalized relative distance ξ between the atoms.

Mechanical Anharmonicity. It is well known that for a mechanically anharmonic oscillator, higher-harmonic absorption transitions appear, which have been observed in many cases [7, 14, 15, 16], although they are rather low in intensity. For a Morse potential, the relative integrated absorption cross sections of these absorption bands can be expressed approximately as

$$\frac{Q_{0n}}{Q_{01}} \approx (n-1)!\, x_{\mathrm{e}}^{n-1}\,. \tag{3.10}$$

These ratios, which are referred to as Rosenthal factors [17], give reasonable agreement with the experimental results for CN^- molecules (as shown for KCl and KBr in Fig. 3.1), but fail almost completely for OH^- and OD^-. In the latter case, even the isotope ratio deviates drastically from the expected value of $1.375^{(n-1)}$.

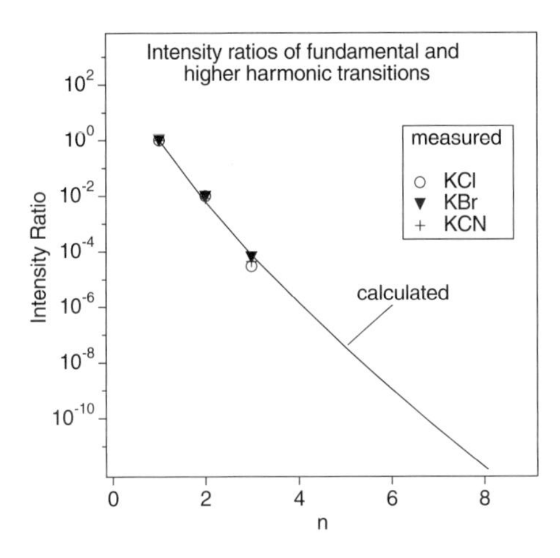

Fig. 3.1. Comparison of the intensity ratio $\frac{Q_{n0}}{Q_{10}}$ observed in VA measurements (from [7]) with values calculated from (3.10) and $x_{\mathrm{e}} = 6 \cdot 10^{-3}$ for the second- and third-harmonic transitions in KCl, KBr, and KCN

Electrical Anharmonicity. To explain the discrepancies in the relative transition strengths of the OH^-/OD^- molecular defects, it is useful to refer back to the Hamiltonian $H' = -p(\xi)E$, which describes the interaction between the electric dipole of the molecule and the electric field of the incident electromagnetic wave. This Hamiltonian leads (as can be shown through application of Fermi's Golden Rule) to a transition probability and to an integrated absorption cross section Q_{0n} which, if we neglect constant factors, can be written as

$$Q_{0n} \propto \overline{\omega}_{0n} \left| \langle n \left| p\left(\xi\right) \right| 0 \rangle \right|^2 . \tag{3.11}$$

Even for a harmonic oscillator nonzero matrix elements are obtained for $n > 1$ if the electric-dipole function is nonlinear. Using the creation and annihilation operators of the harmonic oscillator, it can be shown that

$$Q_{0n} \propto \overline{\omega}_{0n} \frac{\left| \frac{\mathrm{d}^n p(\xi)}{\mathrm{d}^n \xi} \right|^2}{2^n n!} . \tag{3.12}$$

All the nonlinear factors of the electric-dipole function $p\left(\xi\right)$ are collectively referred to as *electrical anharmonicity.* These factors will also modify the relative intensity ratio of an anharmonic oscillator. For a Morse potential, for instance, the ratios for the second- and third-harmonic transitions can be approximated as

$$\frac{I_{02}}{I_{01}} = x_{\mathrm{e}} \left(\left[1 - \frac{\rho''}{\beta} \right]^2 - \frac{6\rho'''}{\beta^2} \right)$$

$$\frac{I_{03}}{I_{01}} = 2x_{\mathrm{e}}^2 \left[1 - \frac{3\rho''}{2\beta} + \frac{\rho'''}{2\beta^2} \right]^2 , \qquad (3.13)$$

where ρ'', ρ''' are relative nonlinearity parameters expressed as the ratios of the second and third derivatives of the electric-dipole function to the first. It is found that ρ'', ρ''' are smaller than 10% for CN^- but can be even larger than 1 for OH^-.

3.2.2 Transition Probabilities for Molecular Ions in a Solid

If we consider a molecule within a dielectric medium with a refractive index n, the expression for the integrated absorption cross sections has to be corrected. Usually, two corrections are considered: (1) the speed of light is reduced, and (2) the effective electric field E_{eff} experienced by the molecule is different from the external electric field. This yields a correction factor $\frac{1}{n} \cdot \left(\frac{E_{\mathrm{eff}}}{E}\right)^2$ for the absorption strength, in which the second term is usually approximated by the Lorenz-Lorentz factor:

$$\frac{E_{\mathrm{eff}}}{E} = \frac{1}{3}\left(n^2 + 1\right) = 1 + \frac{4\pi}{3}\alpha_{\mathrm{total}} = 1 + \frac{4\pi}{3}\left(N_- \alpha_- + N_+ \alpha_+\right) . \qquad (3.14)$$

These expressions show that the correction has its origin in the sum α_{total} of the atomic electronic polarizabilities α_- and α_+ of the N_- and N_+ anions and cations within the sample [18]. This method of simply adding up all contributions is certainly just an approximation, because at any given point within the lattice, the neighbors will be the most important. The tabulated values for α_- and α_+ are, consequently, just sample averages. Still, very reasonable results are obtained for isolated molecules for which the neighbors are the same as the rest of the host ions. In case of the work described in this book, however, in which a close neighbor of the molecule is replaced this approach will fail. In particular, if the electronic defect partner is strongly polarizable, an increase in the measured absorption strength is expected. However, instead of giving a detailed theoretical, quantitative description of the influence of the closest neighbor on the local polarizability, we shall treat a more tractable but instructive example in the following.

For simplicity and for applicability to the F center, we assume that the electronic defect has two electronic states, of gerade (e.g. 1s) and ungerade (e.g. 2p) parity, which are separated by an energy Δ. An electric field in the z direction can mix these states (1s, $2p_z$), resulting in

$$|1\rangle = \frac{1}{\sqrt{1+\varepsilon^2}}\left(|1\mathrm{s}\rangle + \varepsilon\,|2\mathrm{p}_z\rangle\right) ,$$

$$|2\rangle = \frac{1}{\sqrt{1+\varepsilon^2}}\left(|2\mathrm{p}_z\rangle - \varepsilon\,|1\mathrm{s}\rangle\right) ,$$

$$\text{where} \quad \varepsilon = \frac{Ee\,\langle 2\mathrm{p}_z\,|z|\,1\mathrm{s}\rangle}{\Delta} . \tag{3.15}$$

Owing to this mixing the ground state $|1\rangle$ acquires an electric dipole moment $P_{|1\rangle}$, which can be calculated as

$$P_{|1\rangle} = e\,\langle 1\,|z|\,1\rangle = \frac{2\varepsilon}{1+\varepsilon^2} e\,\langle 2\mathrm{p}_z\,|z|\,1\mathrm{s}\rangle . \tag{3.16}$$

For a small mixing parameter ε this induced dipole moment is proportional to ε and (3.15) also to E. In this approximation, one can define an electronic polarizability as

$$\alpha_{\mathrm{el}} = \frac{2}{\Delta} e^2\,\langle 2\mathrm{p}_z\,|z|\,1\mathrm{s}\rangle^2 . \tag{3.17}$$

Inclusion of this polarizability in the effective-field correction yields an additional term (3.14), which for the transition probabilities, leads to a term with an inverse quadratic dependence on the splitting Δ of the relevant states. Furthermore, the influence of α_{el} depends on the distance between the electronic and the molecular defect. This statement becomes more clear if we look at the problem from a slightly different point of view, as follows.

So far, we have treated the electronic defects as part of the polarizable background of the host material. Now we consider them together with the molecule as one vibrating defect. The polarization described by (3.17) causes, in the electric field E_{mol} produced by the dipole moment of the molecular defect, a permanent additional electric dipole (for $p_{\mathrm{mol}}\|z$)

$$p_{\mathrm{add}} = \alpha_{\mathrm{el}} \cdot E_{\mathrm{mol}} = \mathrm{const} \cdot \alpha_{\mathrm{el}} \cdot \frac{p_{\mathrm{CN}}}{R^3} .$$

In a similar way, a change in the molecular dipole $\frac{\mathrm{d}p}{\mathrm{d}\xi}$ causes a change in the additional dipole moment $\frac{\mathrm{d}p_{\mathrm{add}}}{\mathrm{d}\xi}$. This modulation in the mixing of the states results in an additional transition dipole moment, and an absorption strength which is enhanced by

$$I_{\mathrm{add}} \propto \left(\frac{\mathrm{d}p_{\mathrm{add}}}{\mathrm{d}r}\right)^2_{\xi=0} \propto \alpha_{\mathrm{el}}^2 \left(\frac{\mathrm{d}E_{\mathrm{CN}}}{\mathrm{d}r}\right)^2_{\xi=0} \propto \frac{1}{\Delta^2}\frac{1}{R^6} . \tag{3.18}$$

As can be seen from (3.18) the expected enhancement of the absorption crosssection is most pronounced if the splitting between the polarizable electronic levels is small and the two defects are located very close to each other. While this phenomenological picture is able to explain qualitatively the observed enhancement effects, a quantitative description requires a more detailed knowledge of the electronic states involved, the perturbation induced by the molecular defects, and the mutual defect interaction.

3.3 Molecules in Alkali Halides

For further reference, some of the main properties of the OH^- and CN^- molecules as free species and within alkali halides are listed in Table 3.2. The mechanical properties of the oscillator ($\overline{\omega}_e$, x_e) are rather insensitive to the host ions. The main variations occur in the electric properties and in the reorientational behavior.

Table 3.2. Basic properties of CN^- and OH^- molecules as free species and within alkali halide hosts

	Property	OH^-	CN^-
free ion	Size	similar as F^-	similar as Br^-
	Shape	egg-shaped ($r_\parallel \approx 1.6$ Å, $r_\perp \approx 1.2$ Å)	ellipsoid ($r_\parallel \approx 2.2$ Å, $r_\perp \approx 1.8$ Å)
	Interatomic distance	≈ 0.96 Å	≈ 1.10 Å
	Electric dipole moment relative to the center of mass	0.15 eÅ	0.07 eÅ
	fundamental vibrational frequency $\omega_{1\leftrightarrow 0}$	3555 cm^{-1}	not measured
	force constant	8.5×10^5 erg/cm^2	1.6×10^6 erg/cm^2
	anharmonicity x_e	≈ 0.023	not measured
in host crystal	fundamental vibrational frequency $\omega_{1\leftrightarrow 0}$	3570 ... 3740 cm^{-1}	2050 ... 2125 cm^{-1}
	IR absorption oscillator strength f	10^{-4} ... 10^{-2}	$\sim 10^{-2}$
	electric dipole moment relative to the center of mass	0.6 ... 1.25 eÅ	~ 0.07 eÅ
	molecular concentration achieved x	$x \leq 10^{-2}$	$0 \leq x \leq 1$
	orientation	⟨100⟩ in NaCl, KCl, KBr, RbX⟨110⟩ in NaBr, KI⟨111⟩ in CsBr	⟨100⟩ in NaX⟨110⟩ in KX, RbX, CsX, X=Cl, Br, I

The CN^- molecule fulfills in most respects the requirements for a "matrix-isolated" molecule; in most host materials at room temperature, it is able to perform a rotation which is only weakly hindered. The changes for OH^- due to the crystal environment are more drastic resulting in strongly varying dipole moment functions, as can be seen from the wide range of observed oscillator strengths ($\propto \frac{dp}{d\xi}$) and electric dipole moments ($p_{\xi=0}$). A characteristic feature of OH^- is a strong size misfit in most alkali halides studied, which allows translational motion of OH^- between off-center positions within the vacancy, besides rotation. Through interaction with the host ions, the molecule influences the positions of the nearest neighbors at any instance. This "dressing effect" slows down the tunneling reorientation of the molecule because of an increased "effective moment of inertia" [19]. For this reason, OH^- can easily be stabilized and orientationally aligned in off-center positions by an additional dressing caused by defects and impurities.

References

1. J. L. Dunham, Phys. Rev. **35**, 1347 (1930).
2. G. Herzberg, *Spectra of Diatomic Molecules* (Van Nostrand, New York, 1950).
3. P. Morse, Phys. Rev. **34**, 57 (1934).
4. M. Klein, B. Wedding, and M. Levine, Phys. Rev. **180**, 902 (1969).
5. G. Field and W. Sherman, J. Chem. Phys. **47**, 2378 (1966).
6. G. Pandey and D. Shukla, Phys. Rev. B **4**, 4598 (1971).
7. C. P. An, Ph.D. thesis, University of Utah, 1995.
8. M. Huggins and J. Mayer, J. Chem. Phys. **1**, 643 (1933).
9. M. Born and K. Huang, *Dynamical Theory of Crystal Lattices* (Clarendon Press, Oxford, 1966).
10. B. Dick, Phys. Status Solidi B **141**, 61 (1987).
11. L. Bosi and M. Nimis, Nuovo Cimento D **13**, 377 (1991).
12. D. Durand, L. S. do Carmo, and F. Luty, Phys. Rev. B **39**, 6096 (1989).
13. N. H. Rosenbaum, J. C. Owrutsky, L. M. Tack, and R. J. Saykally, J. Chem. Phys. **84**, 5308 (1986).
14. M. Krantz and F. Luty, Phys. Rev. B **37**, 8412 (1988).
15. W. B. Fowler, R. Cappeletti, and E. Colombi, Phys. Rev. B **44**, 2961 (1991).
16. R. Capelletti, W. F. G. Ruani, and L. Kovacs, Defects Amorph. Mater. **16**, 189 (1987).
17. J. Rosenthal, Physics **21**, 281 (1935).
18. J. Tessman, A. Kahn, and W. Shockley, Phys. Rev. **92**, 890 (1953).
19. S. Kapphan, J. Phys. Chem. Solids **35**, 621 (1974).

4 Theoretical Models for E–V Transfer

Soon after the first observation of E–V energy transfer several theoretical models for its description were proposed and compared with the experimental results. The various different approaches will be reviewed and compared in the following. The main focus will be on a critical examination of the statements made by the respective authors about the agreement of their models with experimental data. In addition, the question of what kind of additional experiments would be helpful to further clarify the nature of the transfer will be addressed.

4.1 Common Features

Despite their rather simple appearance, the electronic–molecular defect systems under consideration are not amenable to complete "ab initio" calculations. As a result, theories treating the system as a combined entity are lacking at present. The difficulties arise from the mixed nature of the complex, which contains an F centers or rare-earth ions with a diffuse or complicated electronic structure, a covalent molecule, and an ionic crystal. This makes the application of a single established method (e.g. the density functional or the tight-binding method) essentially impossible. As similar problems arise in other systems combined techniques such as the "embedded cluster method" [1] have recently been developed but have not yet been applied to these defect systems. In the theoretical approaches pursued so far, the partners in the complex are treated as individual weakly interacting defects.

In all these approaches we are dealing with three subsystems; namely electrons, phonons, and molecular vibrations, which undergo a change in their state of excitation during an E–V energy transfer process from an electronically excited defect to an excitation involving the vibrational modes of a molecule. For weak interaction, the contribution of each transition to the total transfer probability can be separated, and each of them contributes a factor (A, B, or C) to the overall probability of the process. If we take as the starting point a situation in which one defect is in an electronic excited state while the molecular defect is still in its vibrational ground state, we are dealing with the following changes in the subsystems:[1]

A: Deexcitation of the electronic subsystem from the electronic excited state Φ_e to the ground state Φ_g, $\Phi_e \to \Phi_g$, giving a factor A.
B: Excitation of lattice phonons $\chi_e^0 \to \chi_g^i$; giving a factor B.
C: Excitation of molecular vibrations $\psi_e^0 \to \psi_g^v$, giving a factor C.

Additionally, energy has to be conserved. This is schematically indicated in Fig. 4.1.

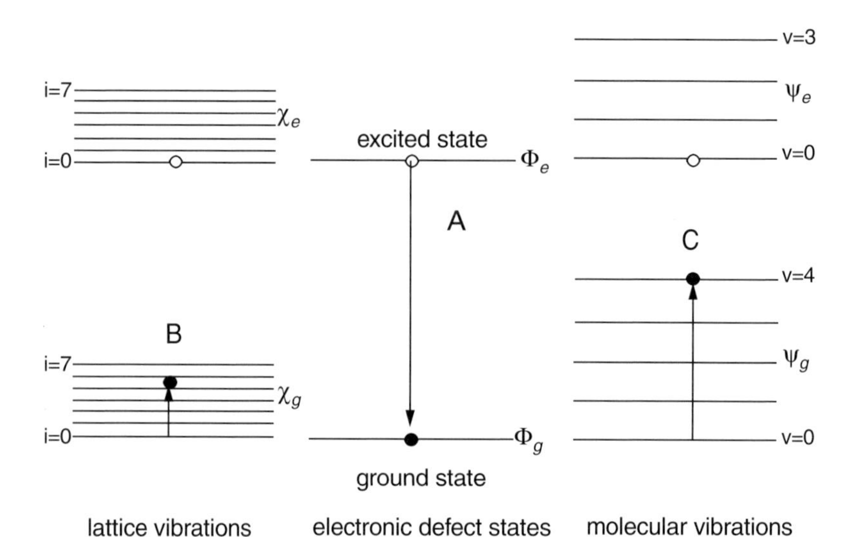

Fig. 4.1. The three subsystems relevant to E–V transfer. *Open circles* describe the state before the E–V transfer and *filled circles* describe the state after it

In taking this point of view, we have employed the commonly used adiabatic (Born–Oppenheimer) approximation which allows us to consider the complete wavefunctions of the initial state $|i\rangle$ and the final state $|f\rangle$ as products of electronic and vibrational wavefunctions: $|i\rangle = \Phi_e \cdot \chi_e^0 \cdot \psi_e^0$ and $|f\rangle = \Phi_g \cdot \chi_g^i \cdot \psi_g^v$. To keep things simple, we represent the different phonon modes by a single "typical" mode.[2]

In all the models described below the transfer probability can be written as

$$W_{e-v} = A \times B \times C\,. \tag{4.1}$$

[1] The subscripts g and e indicate wavefunctions in which the electronic defect is in its ground and excited state, respectively.

[2] If the complete phonon spectrum is considered, χ_{ph}^i has to be replaced by the product over all modes $\prod_j \chi_{j,ph}^i$.

Besides the physical mechanism of the transfer the theories also address the question of the point within the excitation/relaxation cycle where the transfer occurs. Several possibilities have been considered:

(i) During the electronic excitation and/or deexcitation.
(ii) During lattice relaxation.
(iii) In the relaxed exited state.

The various possibilities can be illustrated in the schematic configurational coordinate (CC) diagram shown in Fig. 4.2, proposed by Rong and Luty [2] for the $F_H(CN^-)$ center in CsCl. Despite being very schematic and neglecting many details of the excitation/relaxation cycle, these CC diagrams are a useful tool for visualizing the energetic requirements and can be constructed for other defect systems as well, taking account of the particular electronic states and the electron–phonon coupling (see Figs. 7.6, 11.4, and 12.1). The models which will be reviewed can be divided into two classes:

- *Förster–Dexter (FD)-type models*. The defects interact only through an electric field and their wavefunctions are considered as independent. In

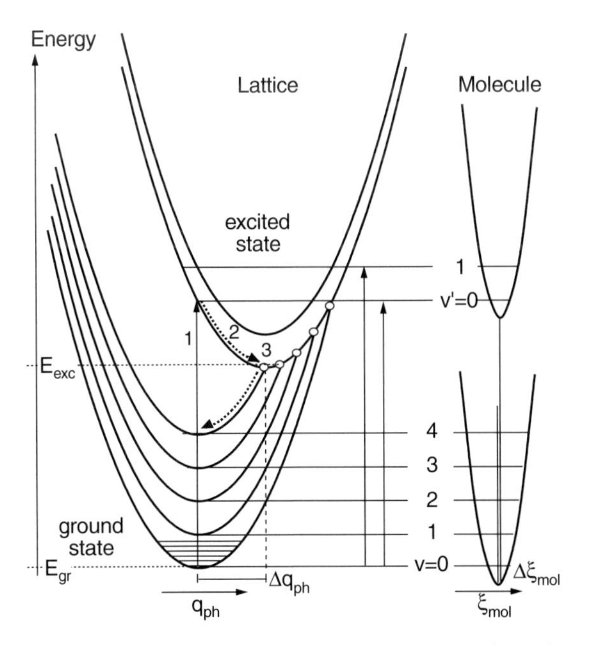

Fig. 4.2. Schematic CC diagram of the $F_H(CN^-)$ center in CsCl, with a configurational coordinate (CC) q_{ph} and parabolic potential curves describing the motion of a typical phonon mode. The energies of excited molecules are indicated by appropriately shifted curves. The points at which the E–V transfer could occur are indicated by numbers (1, 2, 3). On the *right*, the vibrational potentials for the electronic ground and excited states are shown as a function of the normalized relative distance ξ between the two atoms of the diatomic molecule

this approach, theories of transfer between purely electronic or purely vibrational excitations (E–E or V–V, respectively [3, 4, 5]) are adapted to the case of the mixed electronic vibrational (E–V) energy transfer.

- *Electronic–vibrational coupling models*. The vibrational part of the molecular wavefunction is coupled in a similar way to to the electronically excited state of the defect partner as the phonons are. The corresponding E–V transfer theories can be considered as an extension of the well-known theory of the nonradiative decay of defects in solids. Several monographs can be used as a reference for this theory [6, 7, 8, 9]. The vibrational motion of the molecule is included as just an extra mode. In this respect, E–V transfer represents a special case of electron–vibrational coupling. The properties of this *localized* vibrational mode can be studied and quantified much more accurately in terms of eigenfrequencies, anharmonicity, and coupling than in the case of regular phonons. Therefore a clarification of E–V transfer can also contribute to the confirmation or refinement of the "old" theories, and might even reveal inconsistencies.

4.2 Förster–Dexter-Type Models

In this approach, the two defects are treated independently so that the total wavefunctions for the initial and final states can be written as products: $|i\rangle = |i_{\mathrm{mol}}\rangle\,|i_{\mathrm{el}}\rangle$, $|f\rangle = |f_{\mathrm{mol}}\rangle\,|f_{\mathrm{el}}\rangle$ of the individual wavefunctions for the molecule ($|i_{\mathrm{mol}}\rangle$, $|f_{\mathrm{mol}}\rangle$) and the electronic defect ($|i_{\mathrm{el}}\rangle$, $|f_{\mathrm{el}}\rangle$).

In the first attempt to tackle the E–V energy transfer problem, Fowler [10] modified the transfer model of Förster [3] and Dexter [4]. In this theory, which is based on multipolar electric interaction the transfer rate from a donor S (e.g. the F electron in its RES) to an acceptor A (e.g.: a CN^- ion) leading to excitation into the vibrational level v can be written (in the simplest case of dipolar interaction) as

$$W_{\mathrm{e-v}}(v) = \frac{3\,(\hbar c)^4}{4\pi n^4}\,\frac{Q_{\mathrm{mol}}\,(0\rightarrow v)}{R^6}\,k_{\mathrm{e}}^{\mathrm{rad}}\left[\frac{\epsilon}{\kappa^{\frac{1}{2}}\cdot\epsilon_c}\right]^4 \times \int_{-\infty}^{\infty}\frac{F_{\mathrm{el}}(E)F_{\mathrm{mol}}(E)}{E^4}\mathrm{d}E\,. \qquad (4.2)$$

Here F_{el}, F_{mol} are the normalized shape functions of the electronic emission and vibrational absorption, respectively. The expression in parentheses is an effective field correction. Q_{mol} is the integrated absorption cross-section of the vibrational transition $(0 \rightarrow v)$ involved in the transfer. The integral describes the spectral overlap of the donor emission and the acceptor absorption. By evaluating the constants, setting the field correction to 1 and using a delta function $\delta(E - v\hbar\omega_{\mathrm{mol}})$ for the shape function of the vibrational transition (4.2) can be simplified:

$$W_{\mathrm{e-v}} \approx \frac{\mathrm{const}}{R^6} \frac{k_{\mathrm{e}}^{\mathrm{rad}} F_{\mathrm{el}}(\upsilon\hbar\omega_{\mathrm{mol}}) Q_{\mathrm{mol}}}{(\upsilon\hbar\omega_{\mathrm{mol}})^4} \,. \tag{4.3}$$

In this form the factorization (4.1) becomes more obvious. The spectral-overlap condition enforces energy conservation and also includes electron–phonon coupling through the line-shape function $F_{\mathrm{el}}(E)$, which will be discussed in more detail later on.

Using (4.3), Fowler obtained E–V transfer rates by estimating the unknown oscillator strength $Q_{\mathrm{mol}}(\upsilon = 0 \rightarrow 3, 4, 5)$ with the help of the Rosenthal factor [11] which can be calculated from (3.10) using the measured small anharmonicity of CN^- and the oscillator strength of the fundamental transition. The result of his estimate was disappointing, as the rates came out too small by orders of magnitude to compete with the radiative rates of F centers in KCl and CsCl. Although Fowler used values for isolated CN^- defects, the result is not changed drastically if the more recently determined values for the $\mathrm{F_H(CN^-)}$ center [12] are used, instead.

In an attempt to account for the energy transfer of $\mathrm{F_H(OH^-)}$ centers in KCl, Halama et al. [13] deduced a similar relation for the E–V energy transfer caused by dipolar interaction. While their main conclusions have been challenged by Gustin et al. [14] owing to new experimental findings, it is still worthwhile to review their approach, because it reveals the assumptions underlying the Förster–Dexter-type transfer model. Starting with Fermi's Golden Rule, the probability for a transition from $|i\rangle$ to $|f\rangle$ can be written as

$$W_{\mathrm{e-v}} = \frac{2\pi}{\hbar} \sum_f |\langle i \,|H'|\, f\rangle|^2 \times \delta\left(E_f - E_i\right) , \tag{4.4}$$

where E_f and E_i are the total energy of the final and initial states, respectively. We limit our consideration here and in the following to the best-defined situation, that at $\mathrm{T} = 0$ K. If this restriction is released, the sums have to be evaluated over the initial states as well, weighted by their thermal occupation number. For an interaction between two electric dipoles μ_{D}, μ_{A} oriented parallel to one another, we can write

$$H' = \frac{1}{\epsilon R^3} (\mu_{\mathrm{D}} \mu_{\mathrm{A}}) \tag{4.5}$$

and identify μ_{D} as the transition moment from the RES to the electronic ground state and μ_{A} as the expectation value of the molecular dipole operator μ:

$$\mu_{\mathrm{A}} = \langle i_{\mathrm{mol}} \,|\mu|\, f_{\mathrm{mol}}\rangle = \langle i_{\mathrm{mol}} \,|\mu|\, i_{\mathrm{mol}}\rangle \,. \tag{4.6}$$

For the latter assignment, it must be assumed that the electronic wavefunction of the molecule is not changed during the transition. As we shall see, this is a serious restriction on the applicability of the model. However, if we accept this restriction, at least approximately, we can expand μ_{A} around the equilibrium position of the molecular vibration:

$$\mu_{\mathrm{A}} = \mu_{\mathrm{A}}^{0} + \left(\frac{\mathrm{d}\mu_{\mathrm{A}}}{\mathrm{d}\xi_{\mathrm{mol}}}\right)_{\xi_{\mathrm{mol}}=\xi_0} \times \xi_{\mathrm{mol}} + \ldots . \tag{4.7}$$

By neglecting higher derivatives in the expansion, we treat the molecule as a *mechanical* oscillator of two opposite charges, for which the dipole moment changes linearly with their separation. As already discussed in Sect. 3.2, this is again a serious simplification as it is well known for CN^- and OH^- that the details of the electronic distribution play a significant role in the functional dependence of the electric-dipole function (*electric anharmonicity*) [15, 16]. Substituting (4.5–4.7) into (4.4), we obtain for the following transfer rate to a certain vibrational level v:

$$W_{\mathrm{e-v}}(v) = \frac{2\pi}{\hbar\epsilon^2 R^6} \langle\mu_{\mathrm{D}}\rangle^2 \times \left|\left\langle\frac{\mathrm{d}\mu_{\mathrm{A}}}{\mathrm{d}\xi_{\mathrm{mol}}}\right\rangle\right|^2 \left|\langle\psi_{\mathrm{e,g}}^{v}\left|\xi_{\mathrm{mol}}\right|\psi_{\mathrm{e,g}}^{0}\rangle\right|^2 \tag{4.8}$$
$$\times \sum_{i=0}^{\infty}\left(\langle\chi_{\mathrm{g}}^{i}\mid\chi_{\mathrm{e}}^{0}\rangle^2\,\delta\left[E_{\mathrm{e}} - E_{\mathrm{g}} - \hbar\left(i\omega_{\mathrm{ph}} + v\omega_{\mathrm{mol}}\right)\right]\right) .$$

Here E_{e} and E_{g} represent the energies of the electronic excited and ground states. The terms containing μ_{A}^{0} vanish as we consider only cases in which the molecular excitation is changed, and therefore $\mu_{\mathrm{A}}^{0}\cdot\langle\psi_{\mathrm{e,g}}^{v}\mid\psi_{\mathrm{e,g}}^{0}\rangle = 0$; again this only holds as long as the vibrational wavefunctions are not changed during the transition. It is straightforward to show that the two approaches lead to the same results if one considers the following proportionalities:

$$\langle\mu_{\mathrm{D}}\rangle^2 \propto k_{\mathrm{e}}^{\mathrm{rad}} ,$$
$$Q_{\mathrm{mol}} \propto \left|\left\langle\frac{\mathrm{d}\mu_{\mathrm{A}}}{\mathrm{d}q_{\mathrm{CN}}}\right\rangle\right|^2 \left|\langle\psi_{\mathrm{e,g}}^{v}\left|\xi_{\mathrm{mol}}\right|\psi_{\mathrm{e,g}}^{0}\rangle\right|^2$$

and

$$F_{\mathrm{el}}(v\hbar\omega_{\mathrm{mol}}) = \sum_{i=0}^{\infty}\left(\langle\chi_{\mathrm{g}}^{i}\mid\chi_{\mathrm{e}}^{0}\rangle^2\,\delta\left[E_{\mathrm{e}} - E_{\mathrm{g}} - \hbar\left(i\omega_{\mathrm{ph}} + v\omega_{\mathrm{mol}}\right)\right]\right) .$$

Similar FD-type E–V transfer models have also been developed for the non-radiative relaxation of rare-earth and transition metal ions in solutions. The formulas obtained [17, 18, 19] are again of the same form. They differ from (4.3) only in the prefactor to the spectral-overlap integrals. The value of this factor is based on the assumption that the vibrational relaxation is much faster than the energy transfer, a condition which is usually not fulfilled in the systems that we are considering.

4.3 Models for Electronic–Vibrational Coupling

In the FD-type models reviewed so far, the two defects have been regarded as independent, an assumption which may seem to be too crude, especially if the two partners are located on adjacent lattice sites. In this case the wavefunctions of the molecule and the electronic defect cannot be separated anymore. This restriction can be partially lifted if we treat the electronic defect as being coupled to both the phonons *and* the localized molecular vibration. Before reviewing the models based on this assumption, it is instructive to consider first the optical (electric-dipole) transitions for such a coupled defect system. This will help to distinguish between model-dependent and model-independent factors.

4.3.1 Radiative Transitions Including Electronic–Vibrational Coupling

For the center complexes under consideration, it is necessary to distinguish between two types of coupled modes:

- Phonons, represented by a typical phonon with frequency ω_{ph} and coordinate q.
- The localized stretch mode, with frequency ω_{mol} and normalized coordinate ξ.

Using Fermi's Golden Rule with the electronic dipole operator μ as the perturbation, we obtain the following for the radiative transition probability $W_{\text{rad}}\left(E_{\text{em}}\right)$ from an initial state $|i\rangle = \Phi_{\text{e}}\chi_{\text{e}}^{0}\psi_{\text{e}}^{0}$ to a final state $|f\rangle = \Phi_{\text{g}}\chi_{\text{g}}^{i}\psi_{\text{g}}^{v}$ at a certain transition energy E_{em}:

$$
\begin{aligned}
W_{\text{rad}} &= \text{const}\sum_{f}\left|\langle f\left|\mu\right|i\rangle\right|^{2}\times\delta\left(E_{f}-E_{i}-E_{\text{em}}\right)\\
&= \text{const}\left|\mu_{\text{D}}\right|^{2}\sum_{i=0}^{\infty}\left|\langle\chi_{\text{g}}^{i}\mid\chi_{\text{e}}^{0}\rangle\right|^{2}\sum_{v=0}^{\infty}\left|\langle\psi_{\text{g}}^{v}\mid\psi_{\text{g}}^{0}\rangle\right|^{2} \qquad (4.9)\\
&\qquad\times\,\delta\left[E_{\text{e}}-E_{\text{g}}-\hbar\left(i\omega_{\text{ph}}+v\omega_{\text{mol}}\right)-E_{\text{em}}\right]\\
&= \text{const}\left|\mu_{\text{D}}\right|^{2}G_{\text{el}}(E_{\text{em}},\hbar v\omega_{\text{mol}})\,,
\end{aligned}
$$

where $\mu_{\text{D}} = \left\langle\Phi_{\text{exc}}\left(r,\xi^{0},q^{0}\right)\left|\mu\right|\Phi_{\text{gr}}\left(r,\xi^{0},q^{0}\right)\right\rangle$ is the purely electronic transition dipole moment and $G_{\text{el}}(E_{\text{em}},v)$ is the shape function of the electronic luminescence. In going from the first to the second expression in (4.9) the Condon approximation[3] has been applied. Equation (4.9) follows the form of (4.1), and hence radiative transitions can be considered to provide a possibility for energy transfer which would occur during an absorption or emission process in vibrational sidebands with $\Delta v = v \neq 0$. This possibility has been theoretically considered by Dick and Gutowski [20], who found by a quantum chemical method for CsCl that no change in the vibrational potential occurs during the excitation and therefore that only the "zero-vibron" transitions could take place. However, more recent experimental results show that quite significant changes occur (see [21] and Sect. 5.4).

[3] The vibrational coordinates do not change during the transition.

The strength of the sidebands and the shape function are determined by the two Franck–Condon (FC) factors described by the two overlap integrals $\left|\langle\chi_{\rm g}^{i}\mid\chi_{\rm e}^{0}\rangle\right|^{2}$ and $\left|\langle\psi_{\rm g}^{v}\mid\psi_{\rm g}^{0}\rangle\right|^{2}$. For a harmonic oscillator the factors are nonzero for $\Delta v\neq 0$ and $\Delta i\neq 0$ only if the initial and final vibrational states do not have identical sets of eigenstates, i.e. if the vibrational potential is changed. While the FC factors can be calculated numerically for arbitrary potentials, several analytic expressions exist for certain changes of the potential [6]:[4]

- *Displaced undistorted parabolas.* In this case the potential curves are rigidly shifted relative to each other by Δq in the configurational coordinate. This is referred to as "linear coupling". The FC factors can then be expressed as

$$\left|\left\langle\chi_{\rm e}^{n'}\mid\chi_{\rm g}^{n}\right\rangle\right|^{2}=\frac{n'!}{n!}S\exp(-S)L_{n'}^{n-n'}\left(S\right), \tag{4.10}$$

  where $L_{n'}^{n-n'}\left(S\right)$ are Laguerre polynomials. If the initial or final state is vibrationally relaxed ($n'=0$), the expression simplifies to

$$\left|\langle\chi_{\rm e}^{0}\mid\chi_{\rm g}^{n}\rangle\right|^{2}=\frac{1}{n!}S^{n}\exp(-S). \tag{4.11}$$

  In the above expressions, the commonly used Huang–Rhys factor S has been introduced and can be written as $S=\frac{M\omega}{2\hbar}\cdot\Delta q^{2}$.
- *Distorted parabolas.* Another simple case is that the harmonic potentials just have different shapes, and consequently cause different eigenfrequencies ω and ω'. Owing to the symmetry of the harmonic-oscillator wavefunctions, only the FC factors between states with the same parity are nonzero, so that for $n=0$, only the matrix elements with $n'=$ even have to be considered. These matrix elements yield

$$\left|\langle\chi_{\rm e}^{0}\mid\chi_{\rm g}^{n}\rangle\right|^{2}=2\times\frac{1\times3\times5\times\ldots n-1}{2\times4\times\ldots n}\frac{\sqrt{\beta}}{1+\beta}\left(\frac{\beta-1}{\beta+1}\right)^{n}, \tag{4.12}$$

  where $\beta=\frac{\omega_f}{\omega_i}$.
- *Displaced and distorted parabolas.* In this more general case, which is the simplest realistic case of "nonlinear coupling", the FC factors have been calculated by various authors [22, 23, 24, 25, 26, 27] and represented in different ways. For $\beta>1$ (i.e. $\omega_f>\omega_i$) we obtain

$$\left|\langle\chi_{\rm e}^{0}\mid\chi_{\rm g}^{n}\rangle\right|^{2}=\frac{2\sqrt{\beta}{\rm e}^{-S}}{n!2^{n}(\beta+1)}\left(\frac{|\beta-1|}{\beta+1}\right)^{n}\exp\left(S^{2}\frac{\beta-1}{\beta+1}\right)\left|H_{n}\left(z\right)\right|^{2}, \tag{4.13}$$

  where $z=\Delta q\frac{\beta}{\sqrt{\beta^{2}-1}}$ and $H_n\left(z\right)$ are the Hermite polynomials, which can be obtained from the recursion formula $H_{n+1}\left(z\right)=2\cdot z\cdot H_n-2\cdot n\cdot H_{n-1}\left(z\right)$ together with $H_0=1$ and $H_1=2z$.

[4] The relations are given here for the vibronic phonon states χ. For the corresponding relation for the molecular motion, χ has to be replaced by ψ.

The FC factors for the distorted and undistorted cases calculated by means of (4.11) and (4.13) are shown in Fig. 4.3 for small coupling values (which will be relevant later on). For even values of n the values of the FC factors almost completely coincide with those predicted from (4.12). The FC factors for the distorted case show a rather complicated behavior (jumping up and down) and their relative values depend strongly on the particular combination of S and β, but for $n > 3$ these factors are in general larger than those obtained for linear coupling. Although these higher values are desirable for an effective E–V transfer, most of the following theoretical treatments are limited for simplicity to linear coupling.

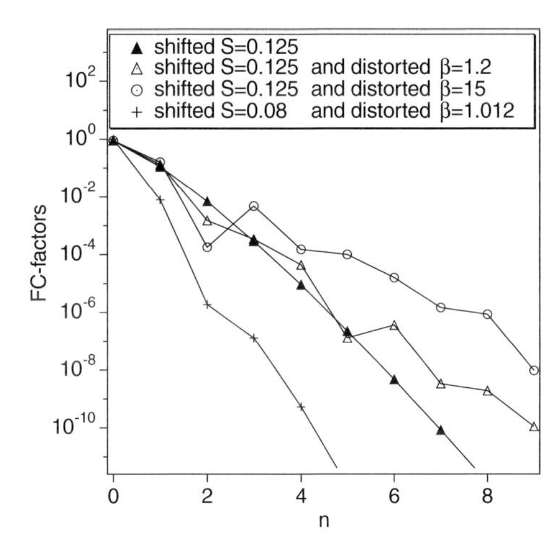

Fig. 4.3. FC factors for different choices of shift $\Delta q = \sqrt{\frac{2\hbar \cdot S}{M\omega}}$ and distortion β of the harmonic vibrational potentials of ground and excited state of the electronic defects

While the cases listed above, which involve only harmonic potentials, are analytically solvable, the more general and realistic case of an anharmonic vibrational potential leads, for all forms usually chosen (e.g. the Morse potential), to complicated analytical expressions [28, 29], which are very awkward to compute [6], so that it is more easier to determine the FC factors approximately by numerical methods using the corresponding wavefunctions [30]. In the case of the Morse potential it is expected that the FC factors will be reduced compared with the harmonic case for large n owing to the decrease of overlap as the Morse potential bends away from the harmonic potential. For smaller values of n, however, an increase is found, especially if the potential and frequency shifts are small. Most notably, the anharmonic contribution is the only one that remains for zero coupling ($S = 0$ and $\beta = 1$), which allows overlap and transitions for $\Delta n \neq 0$ and $\Delta n \neq 1$, respectively. We encountered this situation with the FD models discussed above.

Under the condition of strong linear coupling to the lattice, as is the case for F centers, a further approximation can be applied, which is known as the "short-time approximation". If the condition $S > 1$ is fulfilled, the corresponding term of the line-shape function can be approximated by a Gaussian function. This allows one to write the complete shape function $G_{el}(E, v)$ for the combination of a weakly coupled mode (e.g. CN^-) and a strongly coupled mode (the lattice) as a central Gaussian band $F_{el}(E, v = 0)$ with vibrational sidebands for $v = \pm 1, \pm 2, \ldots$ as follows:

$$\begin{aligned} G_{el}(E,v) &= \sum_{i=0}^{\infty} \left|\langle \chi_g^i \mid \chi_e^0 \rangle\right|^2 \times \sum_{v=0}^{\infty} \left|\langle \psi_g^v \mid \psi_g^0 \rangle\right|^2 \\ &\quad \times \delta\left(E_g - E_e - i\hbar\omega_{ph} - v\hbar\omega_{mol} - E\right) \\ &= \sum_{v=0}^{\infty} \frac{(S_{mol})^v}{v!} e^{-S} F_{el}(E,v) \end{aligned} \tag{4.14}$$

where

$$F_{el}(E,v) = \frac{1}{\hbar\omega_{ph}} (S_{ph})^{-\frac{1}{2}} \exp\left(-\frac{(E_e - E_g - i\hbar\omega_{ph} - v\hbar\omega_{mol} - E)^2}{2 \cdot S_{ph} \cdot (\hbar\omega_{ph})^2}\right) . \tag{4.15}$$

While the width of the main peak is determined by the electron–phonon coupling (which increases with $S_{ph}(\hbar\omega_{ph})^2$), the strength of the first sideband I_1 relative to the main peak I_0 is determined by the coupling to the molecule. For linear coupling the relative strength is simply $\frac{I_1}{I_0} = S_{mol}$. Due to the importance of this relative strength for the E–V transfer models discussed below, the band shape functions for the F center and $F_H(CN^-, OH^-)$ centers have been studied in some detail by West et al. in a series of papers [31, 32, 33]. Unfortunately, the coupling to the molecule has been neglected their papers, so that the results are questionable, as will be shown in Sects. 5.2.2 and 5.4.

4.3.2 Supermolecule Model (Horizontal Tunneling)

In this extension of the well-established theory of nonradiative decay[5] by Halama et al. [13, 34, 35, 36], the starting point is again Fermi's Golden Rule (4.4) using as the perturbation $H' = \frac{\hbar^2}{2}\frac{\delta^2}{\delta q^*}$, which represents the kinetic-energy operator of lattice phonons. For simplicity, just a single "promoting" phonon mode with a normal coordinate q^* is considered. This mode can be different from the accepting modes. While the latter modes are responsible for the dissipation of energy, the promoting modes modulate the electronic wavefunctions in such a way that a transition is induced. In general any given phonon mode can perform both tasks, but for the purposes of calculation it is much easier to treat them separately. H' is the dominant part of the total Hamiltonian which is neglected when the Born–Oppenheimer approximation

[5] The term "supermolecule" was proposed by Halama et al. [34] and will be used here as well, although it is somewhat misleading because the electronic states of the molecule are not considered at all.

is applied. Hence, the underlying origin of the E–V transfer in this model is the "breakdown" of the Born–Oppenheimer approximation. For the total Hamiltonian the Born–Oppenheimer eigenstates are no longer stationary, so that transitions between them can occur. If H' is introduced into (4.4) and the small terms involving second derivatives are neglected, the following expression for the transfer rate for an E–V transition into a vibrationally excited level v at $T = 0$ K is obtained [34]:

$$\begin{aligned} W_{\mathrm{e-v}}(v) &= \frac{2\pi}{\hbar}\hbar^4 \left|\left\langle \Phi_{\mathrm{g}} \left| \frac{\delta}{\delta q^*} \right| \Phi_{\mathrm{e}} \right\rangle\right|^2 \left|\left\langle \chi_{\mathrm{g}}^i \left| \frac{\delta}{\delta q^*} \right| \chi_{\mathrm{e}}^0 \right\rangle\right|^2 \\ &\quad \times \sum_{i=0}^{\infty} \left|\langle \chi_{\mathrm{g}}^i \mid \chi_{\mathrm{e}}^0 \rangle\right|^2 \left|\langle \psi_{\mathrm{g}}^v \mid \psi_{\mathrm{e}}^0 \rangle\right|^2 \\ &\quad \times \delta\left(E_{\mathrm{e}} - E_{\mathrm{g}} - \hbar\left(i\omega_{\mathrm{ph}} + v\omega_{\mathrm{mol}}\right)\right) \\ &= \frac{2\pi}{\hbar} F_{\mathrm{PM}} G_{\mathrm{el}}(0, v) \,. \end{aligned} \tag{4.16}$$

Lines 2 and 3 of (4.16) are identical to the shape function $G_{\mathrm{el}}(0, v)$. The second and third factors in the first line are summarized in the "promoting-mode factor"

$$F_{\mathrm{PM}} = \hbar^2 \left|\left\langle \Phi_{\mathrm{g}} \left| \frac{\delta}{\delta q^*} \right| \Phi_{\mathrm{e}} \right\rangle\right| \left|\left\langle \chi_{\mathrm{g}}^i \left| \frac{\delta}{\delta q^*} \right| \chi_{\mathrm{e}}^0 \right\rangle\right| \,. \tag{4.17}$$

In the case of the linear coupling, $\mathrm{G}_{\mathrm{el}}(0, v)$ can be simplified according to (4.15), using the Huang–Rhys factors (S_{mol}, S_{ph}) to quantify the coupling of the lattice and the molecule to the electronic defect. In their papers Halama et al. assumed F_{PM} to be constant, and obtained a best fit to the relative strengths of their measured Raman lines of $\mathrm{F_H(CN^-)}$ in CsCl using $S_{\mathrm{mol}} = 0.2$ and $S_{\mathrm{ph}} = 33$. However, as mentioned above for the $\mathrm{F_H(OH)}$ center in KCl, some of the assumptions of Halama et al. have been tested and ruled out by more recent experiments [12], so that at the very least their fitting parameters must be changed. We shall come back to this point in the appropriate chapters.

This model of E–V transfer can be easily visualized by use of the diagram in Fig. 4.2. The model corresponds to a "horizontal tunneling" from the $v = 0$ potential curve of the RES to for example. the $v = 3$ potential curve of the electronic ground state. In order for this to occur, the potentials of *both* the phonon *and* the molecular vibrational mode must undergo a change while the electron is making a transition from the RES into the GS.

4.3.3 The Sudden Approximation

A similar but less formal approach to explaining E–V transfer was presented by Pilzer and Fowler [37, 38]. This approach yields similar results to the supermolecule model but gives a more intuitive insight into the problem. The authors assume that the vibrational potential is rapidly altered as the electron performs a transition from the RES to the GS. If this change is taken as linear in the displacement ($a\xi_{\mathrm{mol}}$) the authors find that it corresponds to the linear-coupling case (displaced parabolas). In this "sudden approximation", the molecule finds itself, after the electronic relaxation, in a state which is no longer an eigenstate but a linear combination of new states, with coefficients which can be written as $\left|\left\langle\psi_{\mathrm{g}}^{v} \mid \psi_{\mathrm{e}}^{0}\right\rangle\right|$. Analogously for the system of phonons, the coefficients can be written as $\left|\left\langle\chi_{\mathrm{g}}^{i} \mid \chi_{\mathrm{e}}^{0}\right\rangle\right|$. Enforcing energy conservation "by hand", by choosing proper combinations of v and i the authors find several combinations of coupling parameters S_{mol} and S_{ph} (listed in Table 4.1) which can explain the measured relative E–V transfer efficiency for $F_H(CN^-)$ in CsCl given in [12].

Table 4.1. Combination of values of S_{CN} and S_{ph} that can account for the relative rates given in [12] within the sudden-approximation model of Pilzer and Fowler [37, 38]. The calculated absolute rates $W_{0\to4}$, based on an estimate of the promoting-mode factor, are included as well. Pilzer and Fowler used the following parameters as $E_{\mathrm{e}} - E_{\mathrm{g}} = 1.75\,\mathrm{eV}$, $\hbar\omega_{\mathrm{CN}} = 0.25\,\mathrm{eV}$, and $\hbar\omega_{\mathrm{ph}} = \frac{\hbar\omega_{\mathrm{CN}}}{8} = 0.03125\,\mathrm{eV}$

S_{CN}	S_{ph}	$\left|\left\langle\psi_{\mathrm{g}}^{v} \mid \psi_{\mathrm{e}}^{0}\right\rangle\right|^2 \times \left|\left\langle\chi_{\mathrm{g}}^{i=24} \mid \chi_{\mathrm{e}}^{0}\right\rangle\right|^2$	$W_{0\to4}$ in s^{-1}
6	25	0.010	5×10^{7}
5	24.4	0.0142	7.1×10^{7}
4	23.7	0.016	8×10^{7}
3	22.9	0.013	6.5×10^{7}
2	21.75	0.0065	3.25×10^{7}
1	20	0.00085	4.25×10^{6}
0.2	16	7.8×10^{-7}	3.9×10^{3}

Pilzer and Fowler circumvented the use of Fermi's Golden Rule by directly performing a time-dependent perturbation calculation with a step-function in the vibrational potential. They estimated the promoting-mode factor for the E–V transfer with a simple model based on the motion of the neighboring Cs ions. The electronic wavefunction of the F_H center in the RES was approximated by scaling the function proposed for F centers in NaCl by Fowler [39]. With this estimate Pilzer and Fowler account for the high E–V transfer efficiency within this system only for $S_{\mathrm{mol}} > 2$. If the value given by Halama et al. [34] is taken, however, the transfer rate is too small by orders of magnitude. Moreover, such large coupling constants should be reflected in an apparent sideband structure of the electronic emission. This prediction will be compared with actually measured $F_H(CN^-)$ emission spectra and discussed in Sect. 5.2.2. Apart from their method of enforcing energy conservation "by hand" the sudden-approximation model corresponds essentially to the horizontal-tunneling model in the limit of linear coupling.

4.4 Comparison

The factors A, B, and C obtained from the models considered above are summarized in Table 4.2. Although different ways of enforcing energy conservation have been chosen by different authors, it can be seen that all models considered include a factor B (which reflects the coupling of the electronic defect to the lattice) with the same form and dependence on the coupling strength.

Table 4.2. Factors A , B, and C of the different E–V transfer models. FD_1, Förster–Dexter model of Fowler [10]; FD_2, Förster–Dexter model of Halama et al. [13]; RAD, coupled radiative transition (see Sect. 4.3.1); SM, supermolecule model [34]; SA, sudden approximation [37]. For simplicity, only one phonon mode is considered with identical frequencies for the accepting and promoting modes. For brevity, $F_{\mathrm{el}}(E, v)$ from (4.15) is used. For the SA model, a slightly different separation of the terms was used because the authors ensured energy conservation "by hand" choosing i to be 24

	A	B	C
FD_1	$c_1 \cdot \frac{k_{\mathrm{e}}^{\mathrm{rad}}}{R^6}$	$F_{\mathrm{el}}(0, v)$	$Q_{\mathrm{mol}}\,(0 \rightarrow v)$
FD_2	$c_2 \cdot \frac{\langle \mu_{\mathrm{D}} \rangle^2}{R^6} = c_3 \cdot \frac{k_{\mathrm{e}}^{\mathrm{rad}}}{R^6}$	$F_{\mathrm{el}}(0, v)$	$\left\lvert \left\langle \frac{\mathrm{d}\mu_{\mathrm{A}}}{\mathrm{d}\xi_{\mathrm{mol}}} \right\rangle \right\rvert^2 \left\lvert \left\langle \psi_{\mathrm{g}}^{v}\, \xi_{\mathrm{mol}} \right\rvert \psi_{\mathrm{e}}^{0} \right\rangle \rvert^2$
RAD	$c_4 \cdot \lvert \mu_{\mathrm{D}} \rvert^2$	$F_{\mathrm{el}}(E_{\mathrm{em},v})$	$\left\lvert \left\langle \psi_{\mathrm{g}}^{v} \mid \psi_{\mathrm{e}}^{0} \right\rangle \right\rvert^2$
SM	$\frac{2\pi}{\hbar} \cdot F_{\mathrm{PM}}^2$	$F_{\mathrm{el}}(0, v)$	$\left\lvert \left\langle \psi_{\mathrm{g}}^{v} \mid \psi_{\mathrm{e}}^{0} \right\rangle \right\rvert^2$
SA	$5 \times 10^9 \, \frac{1}{\mathrm{s}}$	$\left\lvert \left\langle \chi_{\mathrm{g}}^{i} \mid \chi_{\mathrm{e}}^{0} \right\rangle \right\rvert^2$	$\left\lvert \left\langle \psi_{\mathrm{g}}^{v} \mid \psi_{\mathrm{e}}^{0} \right\rangle \right\rvert^2$

The factor C, which describes the molecular transition, is different in different models but shows a similar strong dependence on Δv. In order to elucidate this statement, we need to remove the approximation that within a FD-type model the wavefunctions $\psi_{\mathrm{e,g}}$ of the molecular vibration have to be independent of the electronic state. As long as the changes are very small, all steps in the deriviation of (4.8) remain valid within a first-order approximation and we can replace $\psi_{\mathrm{e,g}}$ by ψ_{g} and ψ_{e}. Written in that way, the transition matrix element describes a transition of the molecule under simultaneous change of the electronic state. In a series of papers, Struck and Fonger [26, 27, 40, 41] have shown that the corresponding matrix elements and their thermal averages can be related to each other by recursion formulas. To give an example

$$\begin{aligned} \langle u_n \lvert \xi \rvert v_m \rangle &= \sqrt{\frac{1}{2} m} \cdot \langle u_n \mid v_{m-1} \rangle + \sqrt{\frac{1}{2}\,(m+1)} \cdot \langle u_n \mid v_{m+1} \rangle \\ &\approx \sqrt{\frac{1}{2} m} \cdot \langle u_n \mid v_{m-1} \rangle \ . \end{aligned} \tag{4.18}$$

If, further, we take into account the fact that the wavefunctions of an anharmonic oscillator can be approximated by a sum of harmonic-oscillator wavefunctions, it becomes clear that the factors C found for different models are interrelated through (4.18) and that a change in anharmonicity and vibrational coupling strength will influence all of them. Moreover, the n dependence of the individual contributions themselves is similar. A comparison of the n dependences of the Franck–Condon factors and transition matrix elements of an anharmonic oscillator as depicted in Figs. 4.3 and 3.1 illustrates this point.

For these reasons, it is well justified to consider only one of the models in comparisons of the experimentally determined relative transfer rates with theory. In Sects. 5.2 and 9.10 where the best-defined systems are discussed, we have chosen the FD spectral-overlap model. With use of the measured emission curves, no fitting of the electron–phonon coupling is required. Only the correct v dependence of the factor C has to be found. As was shown in Sect. 3.2, at least the n-dependence of the contribution from anharmonicity can be calculated using the anharmonicity parameters. But many fitting parameters still remain. In that respect, it is easy to obtain a "positive" result in a test of the applicability of the FD model just as has been claimed for other transfer models in [34, 38, 42]. However, *the ability to account for the relative rates is by no means a decisive test for deciding between the different models.* In other words,

> If one model fits the relative rates all others will as well.

Moreover, all changes made to the factors B and C to improve the calculated *absolute* transfer rates will apply to all models in essentially the same way.

A common problem of all models which attempt to calculate the absolute rates is that these rates come out too low when reasonable values of the parameters are used. In the field of nonradiative transitions through electron–phonon coupling, this is a quite common observation related to the difficulty of obtaining the promoting-mode factor. Englman [6] summarized this as follows: "At present this is the weakest and most unreliable point in the theory ... Symptomatic of the difficulties currently besetting the promoting factors is the use of relative decay rates". Although this statement was made already in 1979, this judgment still holds. While it is certainly worthwhile to investigate this theoretical problem further, we shall concentrate in the reminder of this book on the description of the experimental work performed by various groups. We pay special attention to those properties of the defect complexes which enable us to test and narrow down the validity range of the model discussed above. In particular, the quantities and phenomena of interest are

- absolute E–V transfer rates (Sects. 5.2 and 9.8),
- coupling constants (Sects. 5.4 and 9.9),
- the strength of vibrational sidebands (Sects. 5.4, 9.2, and 9.9),

- the properties of the relaxed excited state of the electronic defect (Sects. 5.2, 9.9, and 11.4), and
- changes in the vibrational properties of the molecule.

It will be shown that the most desirable goal, that of to completely describing the coupling and being able to calculate the promoting-mode factors using accurate electronic wavefunctions, cannot be achieved. For that reason, we had to take a more pragmatic approach and use promoting-mode factors from literature that had been determined experimentally [43] or conversely, determine those factors from our measurements. The above list of required information about the system is essentially identical to the list given at the end of Chap. 1. This shows that, besides the general interest in the mutual interaction effects, a detailed knowledge of all these effects is required to explain any single one of them, as demonstrated here for the E–V energy transfer. Furthermore, we shall find, within the diversity of the defect systems studied many arguments that the above models can be considered only as first approximations. Rather than describing the system as two interacting but otherwise independent defects, a treatment as a common entity is certainly more appropriate, but also much more difficult.

4.5 Potential Energy Surfaces

In the first approach to treating the defect system as a single entity Song et al. [44, 45] studied the energy surfaces of $F_H(CN^-)$ centers in KCl and CsCl. Using methods from their work on self-trapped excitons in alkali halides, they combined an electronic wavefunction obtained from the extended-ion method with computer methods of quantum chemistry (the CNDO code) and minimized the total energy for different values of the CN^- bond length ξ and the distance from the molecule Q_2. The latter represents in part the usual configurational coordinate of the lattice. The resulting two-dimensional adiabatic potential-energy surfaces (APES) $V(\xi,\, Q_2)$ exhibit interesting features and also characteristic differences for the two host materials. In CsCl, the APES of the excited and ground states intersect while in KCl there is no such intersection. Generalizing the Dexter–Klick–Russels (DKR)-criterion for nonradiative decay, this observation gives a quite convincing argument as to why the E–V transfer efficiency is so different. Moreover, the APES curves of the excited state show some anisotropy in CsCl depending on whether electronic wavefunctions oriented parallel or perpendicular to the molecular axis are considered. This should be reflected in a dependence of the E–V transfer rates on the excitation energy. Unfortunately, no quantitative results for the transfer rates can be obtained. However, an evaluation of the APES curves shows that drastic changes in the calculated transition dipole moments and bond lengths occur at the intersects of the APES of ground and excited states. This led Song et al. to the conclusion that besides the vibrational coupling (without which no intersection would occur) dipolar interaction like that in the FD models plays a major role in the transfer. This is a Salomon-like judgment in respect of the transfer models reviewed above.

References

1. P. Souchko, A. Shluger, C. R. A. Catlow, and R. Baetzold, Radiat. Eff. Defects Solids **151**, 215 (1999).
2. F. Rong, Y. Yang, and F. Luty, Cryst. Latt. Defects Amorph. Mater. **18**, 1 (1989).
3. T. Förster, Ann. Phys. **2**, 55 (1948).
4. D. Dexter, J. Chem. Phys. **21**, 836 (1953).
5. A. Blumen, S. Lin, and J. Manz, J. Chem. Phys. **69**, 881 (1978).
6. R. Englman, *Non-Radiative Decay of Ions and Molecules in Solids* (North-Holland, Amsterdam, 1979).
7. B. Henderson and G. F. Imbusch, *Optical Spectroscopy of Inorganic Solids, Monographs on the Physics and Chemistry of Materials*, Vol. XVI (Clarendon Press, Oxford, 1989).
8. F. K. Fong, *Theory of Molecular Relaxation: Applications in Chemistry and Biology* (Wiley, New York, 1975).
9. R. Bartram, J. Phys. Chem. Solids **51**, 641 (1990).
10. W. B. Fowler, private communication.
11. J. Rosenthal, Physics **21**, 281 (1935).
12. F. Luty and V. Dierolf, in *Defects in Insulating Materials*, edited by O. Kanert and J.-M. Spaeth (World Scientific, Singapore, 1993), p. 17.
13. G. Halama, K. T. Tsen, S. H. Lin, F. Luty, and J. B. Page, Phys. Rev. B **39**, 13457 (1989).
14. E. Gustin, M. Leblans, A. Bouwen, and D. Schoemaker, Phys. Rev. B **54**, 6977 (1996).
15. W. B. Fowler, R. Cappeletti, and E. Colombi, Phys. Rev. B **44**, 2961 (1991).
16. H.-J. Werner, P. Rosmus, and E.-A. Reinsch, J. Chem. Phys. **79**, 905 (1983).
17. V. Ermolaev and E. Sveshnikova, J. Lumin. **20**, 387 (1979).
18. V. Ermolaev and E. Sveshnikova, Chem. Phys. Lett. **23**, 349 (1973).
19. E. Bodunov and E. Sveshnikova, Opt. Spektrosk. **36**, 340 (1974).
20. B. G. Dick and M. Gutowski, Phys. Status Solidi A **117**, K115 (1990).
21. D. Samiec, H. Stolz, and W. von der Osten, Phys. Rev. B **53**, 8811 (1996).
22. D. Heller, K. Freed, and W. M. Gelbart, J. Chem. Phys. **56**, 2309 (1972).
23. S. Lin and R. Bersohn, J. Chem. Phys. **48**, 2732 (1968).
24. K. Freed and J. Jortner, J. Chem. Phys. **52**, 6272 (1970).
25. C. Manneback, Physica **17**, 1001 (1951).
26. C. W. Struck and W. H. Fonger, J. Lumin. **10**, 1 (1975).
27. C. W. Struck and W. H. Fonger, J. Lumin. **14**, 253 (1976).
28. R. Wallace, Chem. Phys. Lett. **37**, 115 (1976).
29. M. Sage, Chem. Phys. **35**, 375 (1978).
30. D. Samiec, Ph.D. thesis, Universität GH Paderborn, 1997.
31. J. West, S. H. Lin, and K. T. Tsen, J. Chem. Phys. **99**, 7574 (1993).
32. J. West, K. T. Tsen, and S. H. Lin, Phys. Rev. B **50**, 9759 (1994).

33. J. West, K. T. Tsen, and S. H. Lin, Mod. Phys. Lett. B **9**, 1759 (1995).
34. G. Halama, S. H. Lin, K. T. Tsen, F. Luty, and J. B. Page, Phys. Rev. B **41**, 3136 (1990).
35. G. Halama, K. T. Tsen, S. H. Lin, and J. B. Page, Phys. Rev. B **44**, 2040 (1991).
36. G. Halama, Ph.D. thesis, Arizona State University, 1989.
37. S. Pilzer and W. B. Fowler, Mater. Sci. Forum (Proc. ICDIM96) **239–241**, 473 (1997).
38. S. Pilzer, Ph.D. thesis, Lehigh University, 1997.
39. W. B. Fowler, Phys. Rev. **135**, 1725 (1964).
40. W. H. Fonger and C. W. Struck, J. Chem. Phys. **60**, 1994 (1974).
41. C. W. Struck and W. H. Fonger, J. Chem. Phys. **60**, 1988 (1974).
42. J. West, Ph.D. thesis, Arizona State University, 1995.
43. F. D. Matteis, M. Leblans, W. Joosen, and D. Schoemaker, Phys. Rev. B **45**, 10377 (1992).
44. K. Song, L. Chen, P. Tong, H. Yu, and C. Leung, J. Phys.: Condens. Matter **6**, 5657 (1994).
45. L. Song and K. Song, Radiat. Eff. Defects Solids **134**, 405 (1995).

5 $F_H(CN^-)$ Centers

Among the electronic–molecular defect complexes the $F_H(CN^-)$ centers are the ones which have been studied in most detail (starting with the first observation of E–V transfer in 1983 [1, 2]) both experimentally[1] and theoretically (as reviewed in Sect. 4). The main interest of the studies was the very effective E–V transfer especially for Cs halides which leads to intense vibrational luminescence (VL) (see in Fig. 5.3). Even superfluoresence [3] and lasing [4, 5] could be observed. The experimental situation was the topic of several consecutive survey articles by Luty et al. [6, 7, 8, 9], so the description of the properties will be kept brief here.

5.1 Basic Spectroscopic Properties

5.1.1 Overview

The method of forming $F_H(CN^-)$ center follows the well-known procedure described in Chap. 2, consisting of irradiation with light in the F center absorption band at $T = T_2$ for a time which can be optimized for this defect system by monitoring the VL. Little change in the electronic absorption spectra is observed for alkali halides with the fcc crystal structure (e.g. KCl, see Fig. 5.2) in which the F center and the CN^- are $\langle 110 \rangle$ neighbors [10]. For Cs halides (e.g. CsCl, see Fig. 1.1), on the other hand, a clear splitting into two subbands $F_H(1)$ and $F_H(2)$ is found, which represent (at least to a first approximation) transitions polarized parallel ($1s \rightarrow 2p_z$) and perpendicular ($1s \rightarrow 2p_{x,y}$) to the $\langle 100 \rangle$ axis of the center. The spectroscopic data are listed in Table 5.1. Under excitation into one of the F_H bands, a slightly shifted electronic luminescence can be observed for fcc-type samples, while in Cs halides the emission is (almost) completely quenched. Only with extreme care was a weak EL found in CsCl, which will be discussed in Sect. 5.2.2 along with its time dependence.

The most drastic change within the excitation/relaxation cycle of the F electron is the presence of E–V energy transfer and the resulting vibrational

[1] The early experimental work was performed by Y. Yang in F. Luty's group. Owing to Yang's tragic early death many of the results remained unpublished.

luminescence. In all host materials studied vibrational transitions originating from several different excited v-levels can be observed, as can be seen for CsCl in Fig. 5.3. Similarly, the excitation of vibrational levels can also be seen in anti-Stokes Raman measurements [7, 9, 11].

5.1.2 Electronic Transitions

Several attempts [12, 13, 14] have been made to account for the F_H band splitting for the Cs halides host material (listed in Table 5.1). These theoretical investigations were based on absorption data and assumed models in which the center and the molecular axis are aligned parallel along the $\langle 100 \rangle$ axis. The approach of Gash [14], who considered only mixing of the $2p_z$ and 1s states, can easily be ruled out because it cannot explain the relative absorption strengths of the $F_H(1)$ and $F_H(2)$ bands. The calculations of West et al. [12, 13], on the other hand, give much better agreement, but the neglect of electronic–vibrational coupling in the absorption is a serious restriction as we shall see in Sect. 5.4. Moreover, the excited-state wavefunctions obtained by West et al. are not able to explain the polarization dependence of both the anti-Stokes Raman results [15, 16] and the VL. In the latter case, this failure can be demonstrated by the measurement of the polarization dependence of the excitation efficiency (Fig. 5.1). The experimentally determined degree of polarization for the excitation deviates both in spectral position and in absolute value from the calculated degree of polarization, for which it was assumed, as indicated by the schematic picture, that centers excited into their ($1s \rightarrow 2p_z$) and ($1s \rightarrow 2p_{x,y}$) transitions emit parallel ($\|$) and perpendicular($\perp$),respectively, to the polarization of the exciting laser light.

Table 5.1. Electronic transition energies (in eV) of $F_H(CN^-)$ centers in several alkali halides

	F		F_H			
	Absorption	Emission	Absorption			Emission
Host			$F_H(1)$	$F_H(2)$	Δ_{12}	
CsCl	2.18	1.25	1.87	2.14	0.2	1.375
CsBr	1.96	0.91	1.8	1.93	0.13	–
RbCl	2.03	1.07	2.02	2.03	0.02	1.05
KCl	2.3	1.21	2.3	2.24	0.06	1.09

The results from both Raman scattering and VL suggest that the center is not completely $\langle 100 \rangle$ aligned, a point to which we shall come back when we discuss the polarization of the electronic luminescence in CsCl (Sect. 5.5).

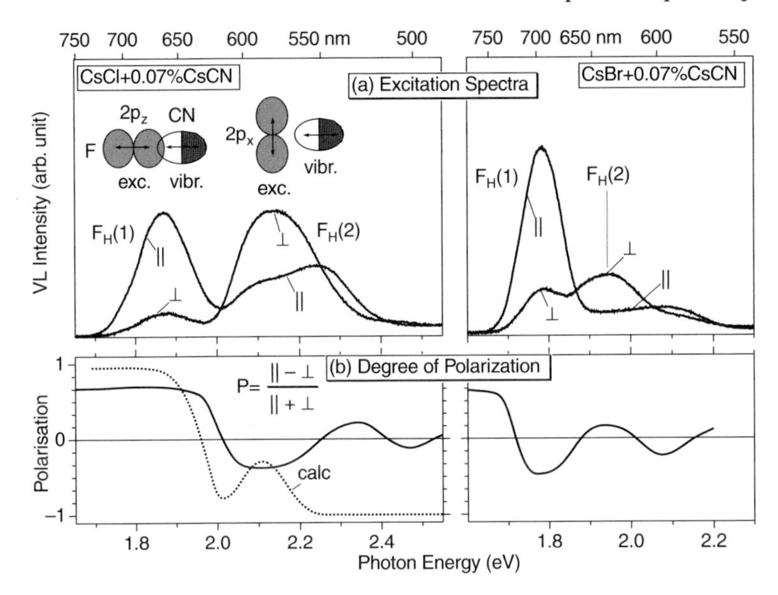

Fig. 5.1. (**a**) Polarization-dependent VL excitation spectra of $F_H(CN^-)$ centers in CsCl and CsBr. (**b**) Degree of polarization. The expected behavior for CsCl using the assumption and fitting parameters of [12], is shown by a *dashed line*

5.1.3 Vibrational Transitions

The data for the vibrational transition energies obtained from VL measurements can interpreted further with 3.3 using for example a method introduced first by Birge and Sponer (see [17]) in which the spectral positions of the anharmonicity-shifted VL lines are plotted versus $\left(v+\frac{1}{2}\right)$. In this way the parameters $\overline{\omega}_e$, $\overline{\omega}_e x_e$, and $\overline{\omega}_e y_e$ can be determined as the fitting parameters of a parabola; they are listed for three alkali halides in Table 5.2 [9]. The parameters $\overline{\omega}_e y_e$ are found to be very small for both isolated and F-center-perturbed CN^-, showing that the Morse potential is a rather good approximation. On the other hand, the relative change of $\overline{\omega}_e y_e$ is very drastic and even includes a change of sign. This term, however, becomes important

Table 5.2. Vibrational parameters of isolated and F-center-perturbed CN^- for several alkali halide hosts, all given in cm^{-1}

	isolated CN^-				$F_H(CN^-)$			
Host	$\overline{\omega}_{1\leftrightarrow 0}$	$\overline{\omega}_e$	$\overline{\omega}_e x_e$	$\overline{\omega}_e y_e$	$\overline{\omega}_{1\leftrightarrow 0}$	$\overline{\omega}_e$	$\overline{\omega}_e x_e$	$\overline{\omega}_e y_e$
CsCl	2079.1	2104	12.45	0.0179	2068.8	2094.02	12.56	−0.0210
CsBr	2067.1	2091.8	12.25	0.016	2060.5	2085.5	12.5	−0.011
RbCl	2079.7	2104.3	12.3	–	2079.1	2103.9	12.4	–

only for very high vibrational excitation levels with large amplitude, which are influenced more by the change in the lattice environment. The difference in $\overline{\omega}_e y_e$ is reflected in the VL spectra by the fact that the energy difference between adjacent vibrational transitions decreases for isolated CN^-, while it increases for $F_H(CN^-)$. If this cubic term is neglected, the Morse potential parameters U_D and β can be calculated and are listed in Table 5.3. Just as for $\overline{\omega}_{CN}$ and $\overline{\omega}_{CN} x_e$ the differences are very small.

Table 5.3. Morse potential parameters U_D, β of isolated and F-center-perturbed CN^- for several alkali halide hosts

	isolated CN^-		$F_H(CN^-)$	
Host	U_D (eV)	β (in Å)	U_D (in eV)	β (in Å)
CsCl	11.02	2.20	10.82	2.21
CsBr	11.12	2.18	10.78	2.20
RbBr	11.16	2.19	11.06	2.20

5.2 Energy Transfer: Relative and Absolute E–V Transfer Rates

The relative E–V energy transfer rates have been determined by time-dependent measurements of the VL on a millisecond timescale [6, 9] and by anti-Stokes Raman measurements [15, 18]. Both techniques show unambiguously that the transfer occurs into high vibrational levels ($v \geq 3$) in the host materials studied. Most notably, for CsBr, the relative rates depend on which F_H absorption band is excited.

Despite all these extensive studies, which date back to before 1992 and have been reviewed by Luty et al. [7, 9], many open questions remained. In particular, many of the parameters, which are required to estimate reliably the E–V transfer rate with the models reviewed in Sect. 4, were not determined. In the following we shall focus on the progress which has been achieved since then. The following aspects will be treated in particular:

- Absolute E–V transfer rates.
- Electronic–vibrational coupling parameters.
- Some more insights into the nature and properties of the relaxed excited state.
- Using the above parameters, the E–V transfer rates will be estimated using the models of Sect. 4.

Although most alkali halide host materials have been investigated further, we shall focus in the following mainly on KCl and CsCl samples which are taken as typical representatives of hosts with an fcc and a simple cubic crystal structure, respectively. Because of this choice, we shall also be able to address the question of why the E–V transfer is so different in its efficiency between these two material classes.

As was pointed out in Sect. 4, the determination of the absolute E–V energy transfer rates is crucial to judging the various models that have been proposed. Owing to the high efficiency and the resulting short transfer times, this task is fairly difficult, especially for the $F_H(CN^-)$ centers in cesium halides. Several techniques have been employed:

- Evaluation of the quantum efficiency of the transfer, which for Cs halides can give only a lower limit ($W_{e-v} > 10^7$ s^{-1}) because of the almost complete absence of EL. The results for KCl and other alkali halides with the NaCl structure will be presented below (Sect. 5.2.1).
- Measurement of the recovery time of the electron into its ground state using a pump–probe technique [19]. The high repetition rate and laser power make the results obtained questionable, however.
- Time-dependent measurements of the vibrational substructure in the electronic transitions which appear under conditions of vibrational excitation [20]. In this detailed study, special care was taken to reduce the repetition rate to a value which guaranteed complete relaxation between pulses. For CsBr, a transfer time of 70 ns was measured. For CsCl, faster time responses were observed but were attributed to interaction effects among excited F centers, which were present in high numbers (50%) owing to the intense laser pulses. For this reason, other methods to determine the transfer rate are required in this host.
- Direct measurements of the onset of VL and the decay of the EL. This technique is the most convincing method because the concerns about high laser power and repetition rate can be ruled out completely. Previously unpublished results [21] will be presented in Sect. 5.2.2 for $F_H(CN^-)$ centers in CsCl.

5.2.1 E–V Transfer Efficiency in KCl

In the first paper on E–V energy transfer, Yang and Luty [1] reported an E–V transfer efficiency for the KCl host material of approximately 3–4%. They based this estimate on measurements of the integrated emission intensities for electronic and vibrational luminescence. However, it was shown (already by the Yang and Luty) that this method is problematic, because the efficiency was found to be dependent on the excitation energy [7]. For this reason, it was crucial to study the excitation-energy dependence of the VL and EL in more detail [21].

The results are shown in Fig. 5.2. Before aggregation of the centers the excitation spectrum of the EL follows closely the F center absorption, as expected. The difference in shape is due to a saturation effect of the emission

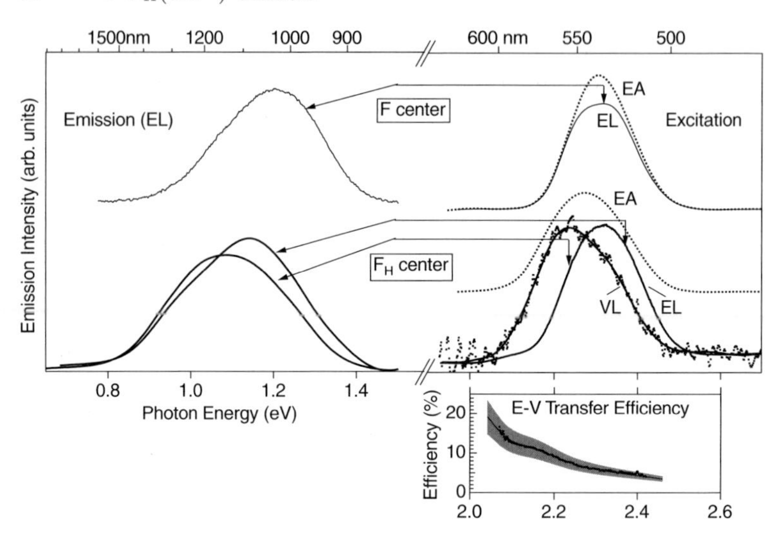

Fig. 5.2. Excitation and emission spectra of EL and VL for F and $F_H(CN^-)$ centers in KCl. For comparison, the corresponding absorption spectra (EA) are shown by *dotted lines*. At the *bottom-right*, the E–V transfer efficiency as a function of excitation energy is depicted

intensity when most of the photons are absorbed. The situation is quite different after the aggregation process. The excitation spectra of the EL and VL are quite different from each other, and neither of them coincides with the F_H absorption, which has been shifted by about 0.4 eV relative to the regular F band. This discrepancy is further reflected in the excitation-energy-dependent EL spectra (Fig. 5.2, left). All these differences are reduced with increasing temperature and vanish for $T > 120$ K. Similarly to what was done by Yang and Luty, the emission intensities can be converted into E–V transfer efficiencies. The result obtained from the excitation spectra of the EL and VL is shown at the bottom of Fig. 5.2. Obviously, excitation on the low-energy side of the absorption leads to a relaxation into the RES, with a lower energy (as can be seen from the emission spectra) and a higher E–V transfer probability. An efficiency of up to 20% can be observed. On the high-energy side, the efficiency is very low. These results indicate that the EA band is not purely homogeneously broadened but must consist of at least two contributions. If we make the assumption that a considerable E–V transfer takes place only for excitation within the lower-energy subband, the excitation spectra of the VL and EL can be used for a decomposition of the EA band into two subbands. The resulting peak energies are included in Table 5.1. Moreover, we have to conclude that relaxation out of the two excited states does not lead to a common relaxed excited state, because differences in the emission spectra and in the E–V transfer efficiency have been observed. The excitation-energy dependence of the relaxation dynamics is further reflected in the time dependence of the EL, which shows an increased decay rate for a lower excitation photon energy. Although other processes may play a role, this observation is completely consistent with an increasing role of the E–V energy transfer relaxation channel.

Similar but smaller effects of this kind are observed for RbCl and KBr. Owing to their importance, the main conclusions of this section are summarized below:

- The broadened F_H absorption consists of at least two distinct subbands which remain unresolved in the absorption measurement.
- Excitation within these subbands leads to different E–V transfer efficiencies and emission spectra, and therefore a *common relaxed excited state has to be excluded.*

5.2.2 E–V Transfer in CsCl: Time-Dependent Measurement of the EL and VL

As has been pointed out before, the determination of the absolute E–V transfer efficiency is particularly difficult for the CsCl host. Owing to the almost 100% efficiency, the rate cannot be deduced from steady-state measurements of EL and VL intensities, and therefore the transfer has to be measured directly from the time dependence of those intensities. The most convincing approach would be to detect both the decay of the EL and the onset of the VL. A comparison of the two observed rates would give not only the transfer rates but also important insight into the question: "At what point within the optical cycle does the transfer occur?" Two problems meant that this approach was not followed from early on:

(i) The electronic luminescence is "totally suppressed", as stated, for example, in [6].
(ii) The onset of vibrational luminescence is very fast, making detection in the IR difficult.

Both problems have now been tackled and solved [21].

Electronic Luminescence in CsCl. The measurement of the time dependence of the EL required that the first statement above had to be challenged and an EL response had to be found. Owing to the weakness of the $F_H(CN^-)$ emission, the main task is to identify it among the unavoidable emission signals from other unwanted F-aggregate centers. For that purpose, we performed very detailed excitation spectroscopy and were able to find an EL response which could be assigned unambiguously: As can be seen at the bottom of Fig. 5.3, an EL signal is found for a spectrum of excitation energies which coincide well with the characteristic $F_H(1)$ absorption. Within the $F_H(2)$ band, the excitation spectra of the VL and EL and the absorption do not completely match, similarly to what was observed for $F_H(CN^-)$ in KCl.

The strength of the EL is at least 40 times weaker, it is shifted to higher energies, and it is considerably narrower compared with the emission band of the regular F center. As shown in the inset of Fig. 5.3, the EL intensity increases as a function of aggregation time simultaneously with an increase of the VL and a decrease of the F center luminescence, giving more evidence that the emission band found is really correlated with the $F_H(CN^-)$ centers.

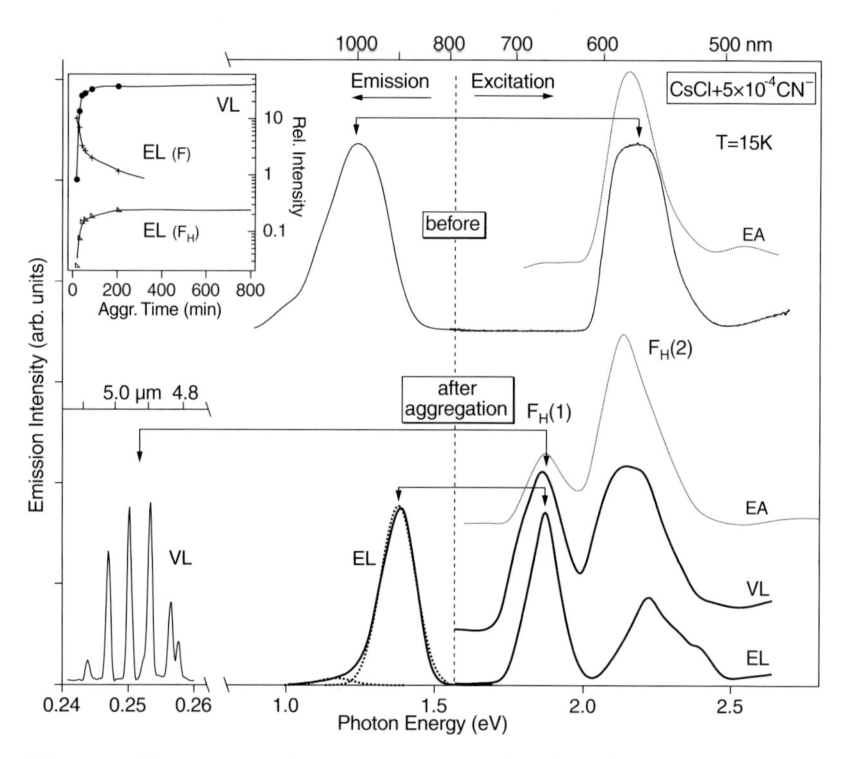

Fig. 5.3. Excitation and emission spectra of $F_H(CN^-)$ centers in CsCl for both EL and VL before and after center aggregation. For comparison the EA spectra are also shown. The *inset* shows the dependence of the VL and EL on the aggregation time. The EL spectrum after aggregation has been fitted (*dotted lines*) according to (4.14)

The spectral shape of the EL can be evaluated using (4.14) and temperature-dependent emission measurements. The spectral width (≈ 0.12 eV) at low temperatures is very similar to that found for the $F_H(1)$ absorption (≈ 0.13 eV) by West et al. [12], suggesting that the coupling of the F electron to the lattice is quite similar for the URES and the RES.

An effective phonon frequency $\omega_{ph}^{eff} = 11$ meV can be found by fitting the T-dependent values of the spectral width to the expression (see Fig. 5.4)

$$\mathrm{FWHM}(T) = \sqrt{2S_{\mathrm{ph}}}\hbar\omega_{\mathrm{ph}}^{\mathrm{eff}}\sqrt{\coth\left(\frac{\hbar\omega^{\mathrm{eff}}}{2kT}\right)}. \tag{5.1}$$

This results in an effective Huang–Rhys factor of $S_{\mathrm{ph}} \approx 21$. The value of the latter, however, is very sensitively related to the fit for $\omega_{\mathrm{ph}}^{\mathrm{eff}}$. For instance, using the value $\omega_{\mathrm{ph}}^{\mathrm{eff}} = 9$ meV found from absorption measurements [12] would result in $S_{\mathrm{ph}} \approx 35$. Owing to this uncertainty, we shall use later on (see Fig. 5.9) the emission spectrum itself to estimate the E–V transfer rates, and not the extracted coupling data.

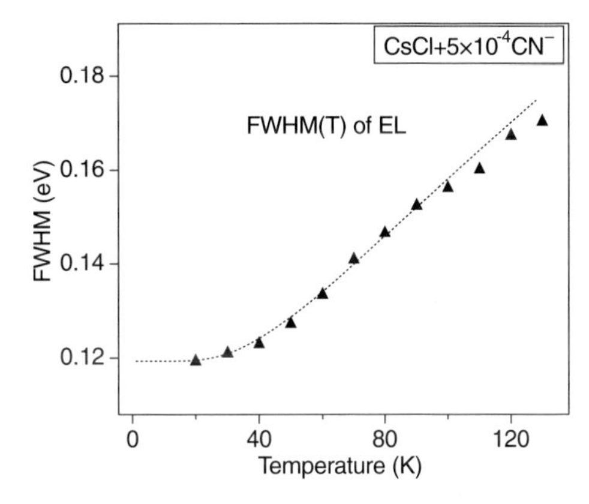

Fig. 5.4. Temperature dependence of the FWHM of the electronic emission originating from $F_H(CN^-)$ centers in CsCl

Turning our attention now to the coupling of the F electron to the CN^- stretchmode, we notice no apparent vibrational sidebands in the EL spectra. By interpreting the low-energy shoulder as a $\Delta v = -1$ sideband (indicated in Fig. 5.3 by a dotted line), we can put a lower limit on the Huang–Rhys factor of $S_{\mathrm{CN}} < 0.02$. This value is much lower than the one considered by Pilzer and Fowler. [22, 23] in their sudden-approximation model described in Sect. 4.3.3. While our value does not exclude the mechanism propsed by Pilzer and Fowler, it seriously restricts the applicability of their results and estimates.

Time Dependence. Once the EL spectrum has been identified, its time dependence can be measured after an excitation pulse from an excimer–dye laser. As shown at the top of Fig. 5.5, the repetition rate of the laser was chosen such that no VL signal could be seen anymore after a delay equal to the time beteen pulses, indicating that the system was completely relaxed. At the bottom of Fig. 5.5, the EL response is depicted along with the excitation pulse, revealing a decay time which is very short and, with a value of 40 ns, about two orders of magnitude shorter than the decay time for the regular EL from nonaggregated F centers.

The question of at which point in the optical cycle the E–V energy transfer occurs can be answered by a measurement of the onset of VL. While this

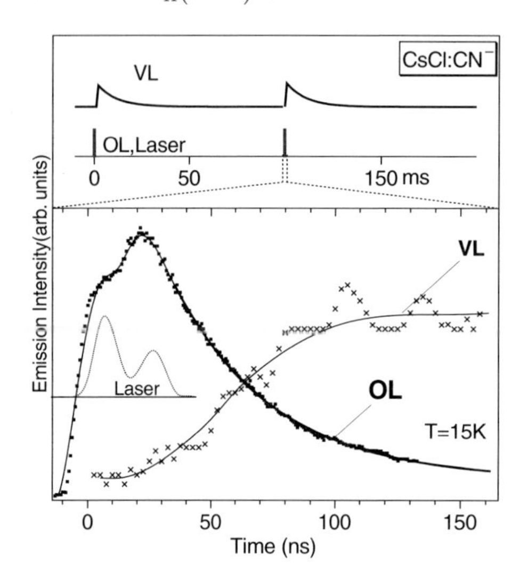

Fig. 5.5. Time dependence of EL and VL of $F_H(CN^-)$ centers in CsCl under pulsed excitation with an excimer–dye laser. The pulse sequence is shown at the *top*

sounds quite easy, the experiments are very difficult. Careful measurements of the VL time response are shown in Fig. 5.5 together with the EL. The onset of VL occurs on the same timescale as the decay of the EL. By fitting the two emission curves using a simple transfer model (described later, in Sect. 9.8), we find good coincidence for nonradiative decay and an E–V transfer time of 40 ns strongly suggesting that both processes are occurring from the same state (i.e. the RES). However, one should keep in mind that the same behavior can also be obtained if an intermediate (time-limiting) state is involved.

5.3 V–E Energy Transfer

Under intense laser excitation, multiple E–V transfer processes can occur, which result in high vibrational excitation of the CN^- molecule. Differently from the case of isolated defects [8], however, it is observed that an "upper limit" exists for this excitation. Only transitions from levels $v \leq 8$ have been observed in VL experiments. The reason for this v-level limit becomes apparent in time-dependent measurements of the VL under intense laser excitation. For CsCl, it was found that the upper vibrational levels have a strongly reduced lifetime (1 μs for $v = 8$) which can be interpreted as the rate of a V–E energy backtransfer from the vibrationally excited molecule to the F center [24]. The vibrational level $v = 8$ coincides energetically very well with the $F_H(1)$ absorption band, making this transfer rather likely from the energetic point of view. The rate of 10^6 s^{-1} appears even more surprising than the rate for the regular E–V transfer because of the high v-number of the vibrational level involved. As can be seen in Figs. 4.3 and 3.1, a transition $v = 8 \rightarrow 0$ is several orders less likely than a transition $v = 0 \rightarrow 4$. On the other hand, we shall see that the parameters for this transfer process are very well defined, because it involves the well-characterized unrelaxed excited state.

5.4 Vibrational Coupling of F Centers to the CN$^-$ Stretchmode

A crucial parameter of all those energy transfer models which attribute the E–V transfer to a breakdown of the Born–Oppenheimer approximation is the coupling of the CN$^-$ molecule to the electronic states of the F centers. Values between 0.2 and 2 have been used to account for the relative transfer efficiency. In principle the molecular–electronic coupling should appear as a vibrational substructure in the absorption and emission spectra but no clear evidence for this coupling has been observed. This fact excludes the large S_{CN} values (see Sect. 5.2.2). The situation is different if transitions are considered for which the CN$^-$ is already excited, because the corresponding Franck–Condon factors $\langle \chi_{\mathrm{CN}}^{v+\Delta v} | \chi_{\mathrm{CN}}^{v} \rangle^2$ increase for higher initial values of v. For that reason, a substructure can be observed in transient absorption detected during the time period in which the molecules are excited after an E–V transfer process. These measurements give the value of the coupling for the unrelaxed excited state. Details of the experimental technique and method of evaluation can be found in [20].

5.4.1 KCl

The experimental results for KCl are shown in Fig. 5.6. From the position and typical shape of the absorption band, it can be estimated that more than 90% of the F centers have been converted to $\mathrm{F_H}$ centers. In the transient-absorption spectra, two sidebands corresponding to transitions with $\Delta v = \pm 1$ appear with the laser excitation and decay at a rate of approximately $(10\ \mathrm{ms})^{-1}$ characteristic of the vibrational relaxation of CN$^-$ in KCl. A negative change is found at the energy of the regular absorption ($\Delta v = 0$). In all the experimental results for KCl, the induced absorption changes are almost an order of magnitude smaller than those obtained under similar experimental conditions for $\mathrm{F_H(CN^-)}$ centers in the cesium halides (discussed below). The separation between the sidebands is $\approx$ 520 meV. This corresponds to twice the eigenfrequency of a CN$^-$ oscillator, which, hence, remains essentially unchanged under electronic excitation. This indicates that the difference in the molecular potential between the ground and excited states is small and justifies the use of the linear-coupling approximation. The change in absorption $\Delta\alpha$ can be expressed as

$$\Delta\alpha = a \sum_{v} \sum_{v'} W_{vv'} N_v g_{vv'}(E) , \tag{5.2}$$

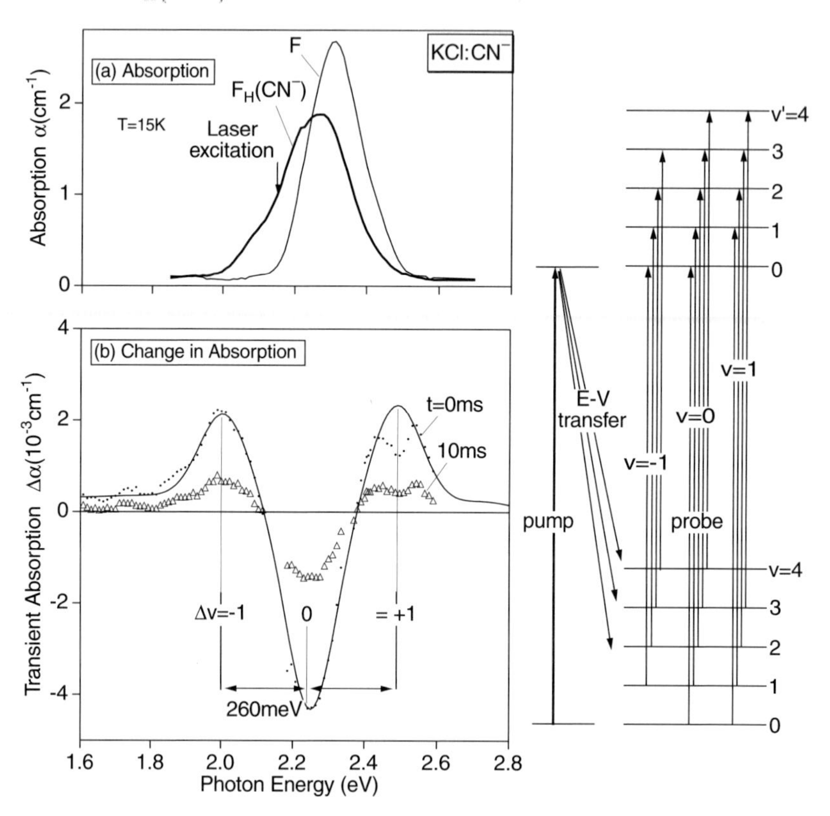

Fig. 5.6. (**a**) Absorption α of F and $F_H(CN^-)$ centers in KCl. (**b**) Transient absorption $\Delta\alpha$ of $F_H(CN^-)$ centers induced by the laser excitation indicated by the *arrow* in (a). The $\Delta\alpha$ spectrum is shown immediately after the laser is switched off and 10 ms later. The *solid line* represents a fit using (5.2)

where N_v is the population of the vibrational level v in the electronic ground state, $W_{vv'}$ is the Franck–Condon factor between level v in the electronic ground state and v' in the electronic excited state, and $g_{vv'}$ is the normalized electronic-absorption shape function. This corresponds to the more general case where unlike in (4.15) vibrationally excited electronic ground states are considered. By applying (4.10), which holds for linear-coupling, $W_{vv'}$ can be written as

$$W_{vv'} = \exp(-S_{\mathrm{CN}})\frac{v!}{v'!}S_{\mathrm{CN}}^{v'-v}[L_v^{v'-v}(S_{\mathrm{CN}})]^2 , \tag{5.3}$$

where L_i^j are Laguerre polynomials. The best fit (indicated by a solid line in Fig. 5.6) to the experimental result is obtained with $S_{\mathrm{CN}} = 0.005$, a value which has, however, an error margin of about 50% owing to the small $\Delta\alpha$ signal.

5.4.2 CsCl and CsBr

Even more detailed information about the electronic–vibrational coupling and the E–V transfer was obtained by Samiec [20] for CsCl and CsBr in his studies of transient absorption, which were carried out over a wide range of timescales (100 ps–1 s) and with variation of both the pump and the probe wavelength. It was possible to fit the large data set consistently for excitation in both the $F_H(1)$ and the $F_H(2)$ absorption bands using a nonlinear model of the coupling of the molecule to the electron in its unrelaxed excited state. The nonlinear coupling is reflected in a shift of the minimum of the vibrational potential, a change in the eigenfrequency, and different anharmonicities. The strength of the coupling obtained is in obvious contradiction to the theoretical prediction of Dick and Gutowski [25]. As can be seen from the values listed in Table 5.4, the nonlinear contribution to the coupling depends strongly on the excited transition. While for $F_H(1)$ excitation the vibrational frequency is decreased by more than 10%, it is increased for $F_H(2)$ excitation. Some consequences of these results are depicted in Fig. 5.7, where the electronic absorption spectra are shown for two vibrational excitations ($v = 0$ and $v = 4$). These spectra have been extracted from the much more complicated experimental data [20] by means of a fitting procedure. With this simplification, the most important results can be seen more clearly: The $F_H(1)$ absorption shows some vibrational substructure even for $v = 0$, and this becomes so dominant for higher excitations that for $v = 4$ the "zero-vibron" transition is totally suppressed. This appreciable vibrational coupling to the molecular stretch mode was neglected by West et al. [12] in their interpretation of the F_H absorption behavior. As one of their implicit fitting parameters was the relative oscillator strength of $F_H(1)$ and $F_H(2)$,

Table 5.4. Changes in eigenfrequency and equilibrium position of CN^- molecules between ground state (GS) and excited states of $F_H(CN^-)$ centers in CsCl, CsBr and KCl. In order to fulfill the condition $\beta \geq 1$ for the applicability of (4.13), the values of β are given either as $\beta = \frac{\omega_{GS}}{\omega_{FH}}$ or as $\beta = \frac{\omega_{FH}}{\omega_{GS}}$ (where ω_{FH} and ω_{GS} are the eigenfrequencies of the molecule when the F center is in its excited and in its ground state, respectively)

Host	state	S_{CN}	ΔQ_0 in Å	$\hbar \cdot \omega_{CN}$ in meV	$\overline{\omega}_{CN}$ in cm^{-1}	β	x_e
KCl	F_H	0.005	0.005	259 ± 5	2089 ± 40	1	?
	GS	–	–	258.8	2087	–	0.006
CsCl	$F_H(1)$	0.12	0.024	215	1734	1.2	0.012
	$F_H(2)$	0.016	–0.009	277	2234	1.07	0.015
	GS	–	–	259.6	2068.8	–	0.006
CsBr	$F_H(1)$	0.005	0.005	247	1992	1.05	0.025
	GS	–	–	258.6	2060.5	–	0.006

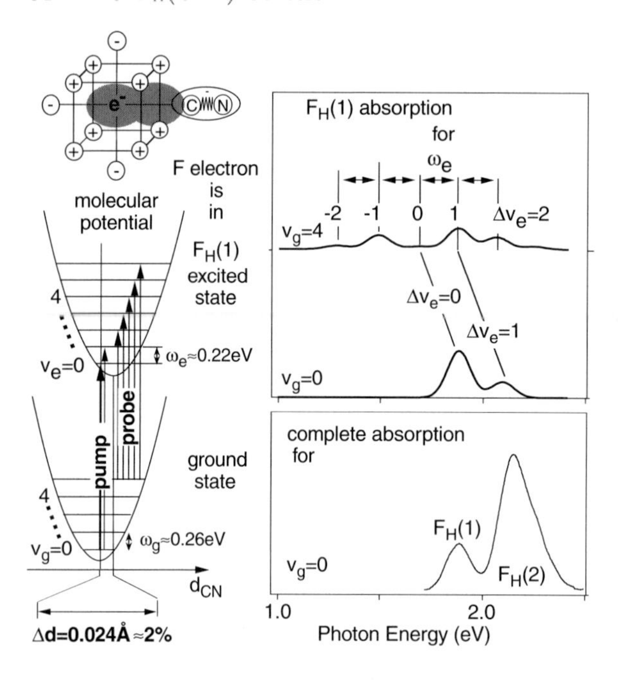

Fig. 5.7. Decomposed absorption spectra originating from a totally relaxed ground state $v_g = 0$ and for a vibrationally excited state ($v_g = 4$) obtained for the $F_H(CN^-)$ center in CsCl under irradiation in the $F_H(1)$ band (the contributions from the $F_H(2)$ band are omitted). On the *left*, a schematic CC diagram is shown, illustrating the nonlinear F-center–CN^- coupling

their results are seriously influenced and are even rendered useless, at least in their quantitative form, by the findings of Samiec et al. [26].

The drastic changes in the vibrational frequency of the CN^- with an excited F electron neighbor cannot be explained by the simple phenomenological picture discussed in Sect. 3.1.2, in which the shift was associated with the repulsive interaction with the neighbors and, hence, the space available for the molecule. In case of $F_H(1)$ excitation, the value of the eigenfrequency is even below the value expected for free CN^- (see Table 5.4). Similar changes are observed only in molecules where the C–N bond is weakened through electron donation from a molecular partner. The antibonding character of the electron-accepting orbital causes a reduction of the vibrational eigenfrequency, a phenomenon new to the field of defects in alkali halides but well known in the chemistry of metal complexes [27, 28]. The expected effect is opposite if one of the electrons is drawn away from the CN^- molecule and its intramolecular triple bond toward a bonding partner. In our case, this can occur when the F center is in a $p_{x,y}$-type state and the electron distribution, from the viewpoint of the molecule is reduced in the center of the anion vacancy, giving rise to an attractive potential. We shall come back to this kind of model in Chap. 9 when we discuss the spectral shifts observed for Yb^{2+}–CN^- complexes.

Comparison. If we compare the data of Table 5.4 for different host materials, a clear correlation between the E–V transfer efficiency and the vibrational coupling can be found. This is especially true for the nonlinear contribution expressed by the change in eigenfrequency. The shift in the average bond length expressed by S_{CN} seems to have less influence, indicating that a restriction to linear coupling will not allow us to explain the host dependence of the transfer rates.

5.5 The Nature of the Relaxed Excited State

The most important requirement for a description of the E–V energy transfer rate, but also the main difficulty, is an understanding of the relaxed excited state from which the transition occurs. Unfortunately, for all alkali halides the conclusion of G. Baldacchini [29] in his review article on the RES of 1992, which he took from an article published ten years before, is still more or less valid: "Evidently a complete model for the RES that provides a completely consistent quantitative explanation for all data has not been achieved."

Owing to the amount of available experimental data, the situation is best characterized for the F center in KCl. For this and other K-halide hosts at least, it is now widely accepted that the RES is a fairly diffuse state in which a considerable amount of mixing between the 2s and 2p states is present owing to coupling to the lattice. The amount of this mixing and the timescale on which the crossover between the 2p and 2s potential curves occurs have still not been fully investigated. For instance, detailed measurements on the loss of polarization memory indicate that the lattice relaxation is finished before the 2p/2s crossover [30]. For our E–V transfer problem, this could have the consequence that the energy transfer may take place also out of a 2p-type state before the electron finally reaches the 2s state. Much less is known for the Cs halides.

The state of knowledge about the RES of the $F_H(CN^-)$ center is even worse, both because fewer details are accessible experimentally and because the situation is certainly more complicated. To underline the latter statement, we recall that we have already had to assume the existence of two different RESs, which are populated in a way that depends on the excitation energy and the resulting way of relaxation.However, we can retrieve, at least some information about the RES from an analysis of the measured electronic luminescence. Besides a shift to higher energies, in CsCl we observed a reduction in the spectral width compared with the F center, suggesting that, as for the $F_H(1)$ band, the coupling to the lattice is reduced. The very complicated and so far not understood microscopic structure of the RES becomes apparent if polarization-dependent excitation spectra are considered (Fig. 5.8). In a result that is completely unexpected if we use the $\langle 100 \rangle$ center model, which in a first approximation was able to explain the absorption behavior, one observes not a $\langle 100 \rangle$ polarization, but a $\langle 110 \rangle$ dichroic polarization effect for $\langle 110 \rangle$ excitation.

This observation can only be explained if we assume that even in the ground state, the defect system is not perfectly aligned along the $\langle 100 \rangle$ axis

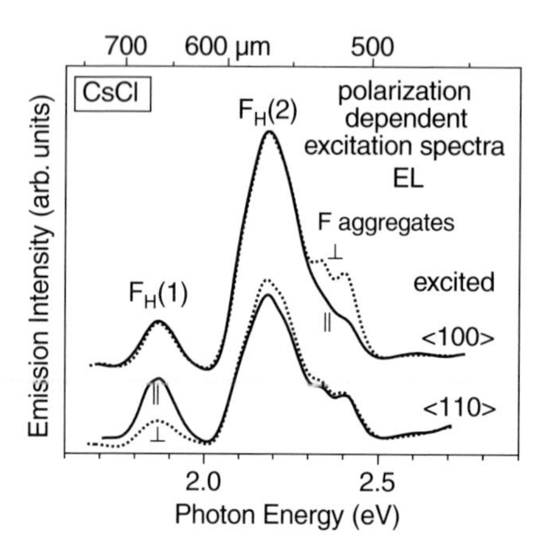

Fig. 5.8. Polarization dependence of electronic luminescence of $F_H(CN^-)$ centers in CsCl under excitation with ⟨100⟩ and ⟨110⟩ polarized light, with detection parallel (*solid line*) and perpendicular (*dotted line*) to the excitation

(as suspected already on the basis of VL and ASRR [15] measurements) and that preferential reorientation of the defect system into a ⟨110⟩- or ⟨111⟩- oriented (probably off-center) arrangement takes place during the excitation–relaxation cycle. While details cannot be extracted from the experimental data, it is certainly conceivable that several local orientational minima exist, which may or may not be reached depending on the excitation channel. On the basis of the experimental findings, one needs to assume two RESs with different E–V transfer properties in CsBr. Similarly, in KCl, the clear differences in the excitation spectra for EL and VL could be explained if E–V transfer is possible for only one of the RESs and not for the other.

The complicated nature and the strong interaction effects occurring during the optical cycle become even more apparent in MCDA measurements, which show that, unlike the case for the F center, the magnetic spin memory is completely lost for the $F_H(CN^-)$ center [31]. In summary, one is certainly dealing with considerable uncertainty and many open questions, which are transferred to the E–V transfer problem whenever the RES is involved.

5.6 Putting It All Together: Comparing E–V Transfer Rates with Theoretical Models

As outlined in Chap. 4, comparison of the measured E–V transfer rates with theoretical models requires a knowledge of several important quantities, which now have been determined, at least for the $F_H(CN^-)$ center in CsCl:

- The electron–phonon coupling ($S_{\mathrm{ph}} \approx 21$), which is included in the shape of the emission spectrum.
- The electronic–vibrational coupling, which we were able to limit to $S_{\mathrm{CN}} < 0.02$ for the RES owing to the absence of vibrational sidebands in the emis-

sion. For the unrelaxed excited state the value for the nonlinear coupling determined in transition absorption measurements can be used (Table 5.4).

- The absolute transfer rates: $W_{E-V} \approx 2.5 \times 10^7$ s^{-1}.
- The relative E–V transfer rates from Ref. [9].
- The anharmonicity constant $x_e = 6 \times 10^{-3}$.

While the above parameters are not well defined for the E–V transfer that occurs in the relaxed excited state, the V–E energy transfer from a CN^- ion in its $v = 8$ excited level to the unrelaxed excited state of an F center all required parameters are available with the exception of the factor A which was introduced in Chap. 4.

5.6.1 FD Model

The parameter A can be determined within an FD-type model, and one finds rates which are several orders of magnitude too low if just the $v = 8 \rightarrow 0$ transition probability due to anharmonicity is used. If, additionally, the change in bond length and the shift in eigenfrequency during the transition are included, higher rates are obtained, but these are still at least one order of magnitude too low. This applies for both the V–E and the E–V energy transfer rates. For the latter, however, a coincidence of the transfer rate can in principle be achieved by adjusting the unknown transition probability of the $v = 0 \rightarrow 1$ absorption for a $F_H(CN^-)$ center in its RES. The order of magnitude which can be expected for such an enhancement will be discussed in Chap. 6.

5.6.2 Horizontal-Tunneling Model

Within the horizontal-tunneling model, the promoting-mode factor F_{pm}, which determines the factor A, has not been obtained (so far), and therefore we have choosen another approach. We have used our measured data to calculate the promoting-mode factor required to account for the V–E energy transfer in CsCl. We find a value of $F_{pm} \approx$ 5–100 meV, which agrees quite well with the values estimated for large molecules [32] and for the F center in NaBr [33]. As can be seen from Fig. 4.3, the Franck–Condon factors for the nonlinear-coupling case are very sensitive to the particular values of Δq and β used, so that the determined F_{pm} has a considerable uncertainty. Moreover, owing to the drastic changes of the electronic state during configurational relaxation, F_{pm} is certainly not expected to be identical for the URES and RES. Nevertheless, if we neglect these concerns, the absolute rate of energy transfer from the RES to the CN^- stretch mode can be calculated, and rates comparable to the measured values can be obtained quite easily. For instance, in the case of $F_H(1)$ excitation in CsCl, a rate $W_{E-V} = 10^7$ s^{-1} is found for the following parameters:

- transition into $v = 4$;
- $A = 6 \times 10^{12}$ meV/s, using $F_{pm} = 5$ meV;
- $B = F_{el}(v = 4) \approx 0.1\ meV^{-1}$, using the measured EL spectra;
- $C = 2 \times 10^{-5}$, using (4.13) with the coupling parameters of the URES listed in Table 5.4.

Despite the numerous assumptions the correct order of magnitude for the transfer rates is obtained. Similarly, the measured E–V transfer rates can be accounted for when the vibrational coupling parameters are smaller, as found in other hosts and under other excitation conditions (KCl [34], CsBr [20], and $F_H(2)$ excitation in CsCl [20]). The only difference is that somewhat higher promoting-mode factors and a better spectral overlap of the emission with the vibrational energy have to be assumed. All this shows that the theory is consistent within itself. *This fact, however, should not be mistaken for a complete solution of the* E–V *transfer problem for the* $F_H(CN^-)$ *centers in alkali halides.* In order to achieve this goal, the promoting-mode factor for the RES needs to be calculated, requiring detailed knowledge about the nature of this state. In an initial semiclassical approach by Pilzer, who estimated the promoting force from the odd-parity vibrational modes of the closest, heavy Cs ions, a value for F_{pm} was found for which the transfer rates come out too low if reasonable values of the coupling constants S_{CN} are used [23].

5.6.3 Relative Transfer Rates

Less problematic is the test for the relative E–V transfer rates into the different levels. Using the measured F_H center emission spectra for determination of the electron-phonon coupling and the vibrational transition energies, the rates can well be fitted (see Fig. 5.9) if for the v-level-dependence of the factor C either those determined from anharmonicity (see Fig. 3.1) or those from

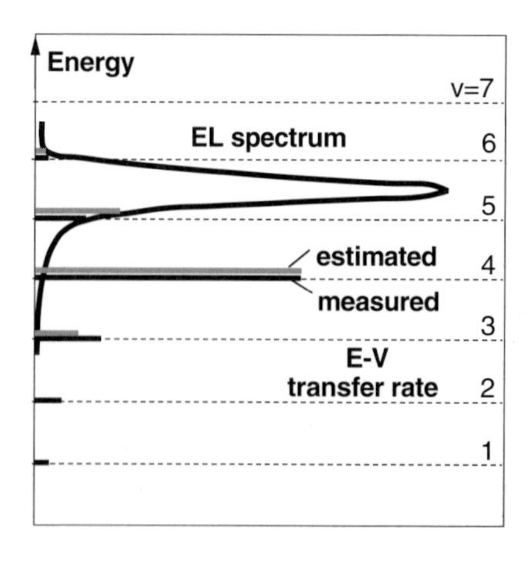

Fig. 5.9. Relative E–V transfer rates (*black bars*) for $F_H(CN^-)$ in CsCl and comparison with values expected from transfer models (*gray bars*). In order to visualize the spectral overlap the electronic emission spectrum (*solid line*) and the position of the vibrational-absorption transitions $v \to 0$ (*dashed lines*) have been drawn on a vertical energy scale

the nonlinear coupling (see Fig. 4.3) are taken. No decision between transfer models is possible on this basis. This problem has already been mentioned above in Chapt. 4.

References

1. Y. Yang and F. Luty, Phys. Rev. Lett. **51**, 419 (1983).
2. Y. Yang, W. von der Osten, and F. Luty, Phys. Rev. B **32**, 2724 (1985).
3. Y. Yang and F. Luty, J. Lumin. **40&41**, 565 (1988).
4. W. Gellermann, Y. Yang, and F. Luty, Opt. Commun. **57**, 196 (1986).
5. W. Gellermann and F. Luty, Opt. Commun. **72**, 214 (1987).
6. F. Luty, *Excited State Spectrosc. in Solids* (XCVI Corso, Soc. Italiana di Fisica, Bologna, 1987).
7. F. Rong, Y. Yang, and F. Luty, Cryst. Latt. Def. and Amorph. Mat. **18**, 1 (1989).
8. V. Dierolf and F. Luty, Rev. of Solid State Sc. **4**, 479 (1990).
9. F. Luty and V. Dierolf, in *Defects in Insulating Materials*, edited by O. Kanert and J.-M. Spaeth (World Scientific, Singapore, 1993), p. 17.
10. T. Pawlik, R. Bungenstock, J.-M. Spaeth, and F. Luty, Radiat. Eff. and Defects Solids **134**, 465 (1195).
11. V. Dierolf, Ph.D. thesis, University of Utah, 1992.
12. J. West, K. T. Tsen, and S. H. Lin, Phys. Rev. B **50**, 9759 (1994).
13. J. West, K. T. Tsen, and S. H. Lin, Mod. Phys. Lett. B **9**, 1759 (1995).
14. P. W. Gash, Phys. Rev. B **34**, 5691 (1986).
15. G. Cachei, H. Stolz, W. von der Osten, and F. Luty, J. Phys.: Condens. Matter **1**, 3239 (1989).
16. R. Albrecht, H. Stolz, and W. von der Osten, J. Phys.: Condens. Matter **4**, 9269 (1992).
17. P. A. Gorry, *Basic Molecular Spectroscopy* (Butterworths, Kent, 1985).
18. K. T. Tsen, G. Halama, and F. Luty, Phys. Rev. B **36**, 9247 (1987).
19. D.-J. Jang and J. Lee, Solid State Commun. **94**, 539 (1995).
20. D. Samiec, Ph.D. thesis, Universität GH Paderborn, 1997.
21. F. Luty and V. Dierolf (unpublished).
22. S. Pilzer and W. B. Fowler, Mater. Sci. Forum (Proc. ICDIM96) **239–241**, 473 (1997).
23. S. Pilzer, Ph.D. thesis, Lehigh University, 1997.
24. V. Dierolf and F. Luty, in *Defects in Insulating Materials*, edited by O. Kanert and J.-M. Spaeth (World Scientific, Singapore, 1993), p. 559.
25. B. G. Dick and M. Gutowski, Phys. Status Solidi A **117**, K115 (1990).
26. D. Samiec, H. Stolz, and W. von der Osten, Phys. Rev. B **53**, 8811 (1996).
27. L. Jones, Inorg. Chem. **2**, 777 (1963).
28. K. Nakamoto, *Infrared and Raman Spectra of Inorganic and Coordination Compounds*, 4th ed. (Wiley, New York, 1986).
29. G. Baldacchini, in *Optical Properties of Excited States in Solids*, edited by B. D. Bartolo (Plenum, New York, 1992), pp. 255–301.
30. N. Akiyama, F. Nakahara, and H. Ohkura, Radiar. Eff. Defects Solids **135**, 345 (1995).

31. G. Baldacchini, private communication.
32. R. Englman and J. Jortner, Mol. Phys. **18**, 145 (1970).
33. F. D. Matteis, M. Leblans, W. Joosen, and D. Schoemaker, Phys. Rev. B **45**, 10377 (1992).
34. V. Dierolf, J. Hoidis, D. Samiec, and W. von der Osten, Radiat. Eff. Defects Solids **149**, 381 (1999).

6 CN^- Next to an Anion Vacancy Occupied by No Electron or Two Electrons

The possibility to photoionize $F_H(CN^-)$ centers makes it possible to create further kinds of centers, in which the CN^- molecule is located next to an unoccupied anion site or an anion site occupied by two electrons. Investigation[1] of these centers allows one to study the influence that a change in electron density has on the molecule, thereby simulating the effects occurring during the excitation/relaxation cycle of a $F_H(CN^-)$. By comparison, a "feeling" for the interaction effects in the RES of the F electron can be obtained.

6.1 Background

Although a detailed description of the RES is not possible for the $F_H(CN^-)$ center, as we have seen in the previous chapter, the physical properties of this shallow, unstable state make photoionized F-type centers very versatile in terms of possible subsequent processes, some of which we have already considered (EL and E–V transfer). At slightly elevated temperatures T_1, the electron can, in competition with EL and E–V transfer, become thermally transferred from the RES into the conduction band (leaving behind an anion vacancy, denoted by F^{ion}), and after conduction can become trapped at another center. These processes have been very well studied for the isolated F center [1] and occur for $F_H(CN^-)$ as well. The possibility of photoionizing the $F_H(CN^-)$ centers opens up the opportunity to study the behavior of the molecular defect in the vicinity of an empty anion site (F^{ion}–CN^-) or an doubly occupied site (F'–CN^-). In this way, the influence of the electron distribution on the properties of CN^- can be studied. By considering the series of defects with 0, 1, and 2 electrons as neighbors a feeling for the order of magnitude of the changes to be expected in the RES can be obtained (Fig. 6.1). At least some qualitative insights are expected, because the RES bears some resemblance, in its delocalization of the electron distribution, to the F'–CN^- center as schematically indicated in Fig. 6.1. As seen in the previous chapter, an increase of the absorption cross section and/or

[1] Many of the experimental results and interpretations presented in this chapter are the product of a collaboration with Professor F. Luty at the University of Utah and mostly have not been published previously.

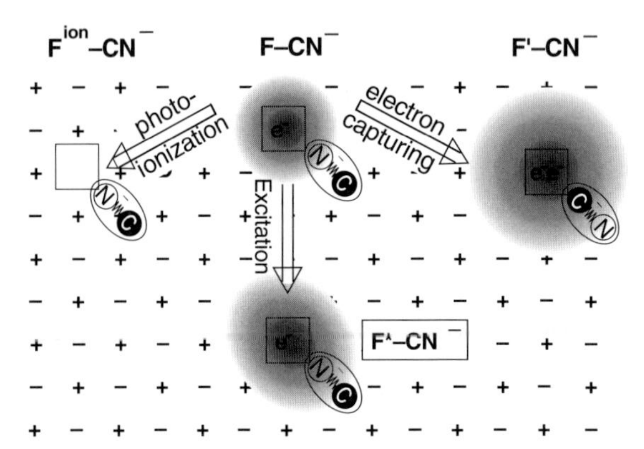

Fig. 6.1. Schematic illustration of changes in electron distribution occurring during excitation, ionization and capture of an electron

the anharmonicity may allow on to obtain quantitative agreement between experiment and an FD-type E–V transfer model.

Two preconditions for this investigation were the ability to photoionize the F_H center at a temperature lower than its dissociation temperature (≈ 230 K for NaCl-type and ≈ 180 K for CsCl-type hosts) and that the electron can be captured by an F_H center. Both these conditions are fulfilled for CsBr and KCl for which results will be presented in the following. In CsCl and RbCl, photoionization but no electron trapping is observed, which results overall in a very inefficient bleaching of the F_H electronic absorption band. While the latter is unfortunate in terms of the present investigations, it is the reason why a vibrational laser based on $F_H(CN^-)$ centers in CsCl [2] can be operated up to higher temperatures than can the $F_H(CN^-)$ laser in CsBr [3]. By the same argument the RbCl host could be – despite the lower E–V transfer efficiency – a further prime candidate for laser experiments.

6.2 Experimental Results

Photoionization of the F_H center is reflected in the electronic absorption by a reduction of the F-type absorption band and the occurrence of a broad EA band, which is very similar to the absorption known for F' centers. For this reason the selectivity is limited in EA measurements and therefore we concentrate in the following on the VA. The best results for all hosts studied were obtained for CsBr, for which photoionization takes place even at $T < 60$ K resulting in VA spectra, as shown in Fig. 6.2. Two additional lines with characteristically different spectral widths appear close together shifted to the low-energy side relative to the isolated and the F-center-perturbed CN$^-$ molecule. These bands and their relation become clearer if the difference spectra before and after ionization are examined. The following observations can be made:

- The integrated VA signal for the centers created is larger than the corresponding signal destroyed by photoionization.
- The relative strengths of the two new lines depend on the wavelength used for photoionization.
- By using polarized light for the photoionization, the lines can be assigned as indicated in Fig. 6.2.
- When the temperature is varied, the VA spectra of the F^{ion}–CN^- center are hardly changed in position and width, indicating a weak coupling to the lattice and strong orientational alignment. The VA from the F'–CN^- center, on the other hand, increases in width and shifts to higher energy. The latter trend is opposite to what is expected as a result of the lattice expansion. The direction of the shift towards the isolated defect reflects a decrease in the interaction that causes the spectral shift.

On the basis of these characteristic properties, it is possible to identify also the new spectral lines that occur in KCl, as shown in Fig. 6.3. In comparison with the isolated CN^- molecule with its librational sidebands, the new features arising from the F^{ion}–CN^- and F'–CN^- centers are barely visible; but again become clear in the difference spectra. Even the VA from F–CN^-, which is usually totally obscured, can be found in this way as a negative signal. In RbCl and CsCl, only one new feature could be observed, which was assigned to F^{ion}–CN^-center owing to its small width. All spectroscopic data for this defect system, for all hosts studied are collected in Table 6.1. The transitions from F'–CN^- and F^{ion}–CN^- centers can also be observed in VL measurements, in addition to the VA measurements, most likely because the centers are created at already vibrationally excited F–CN^- center sites.

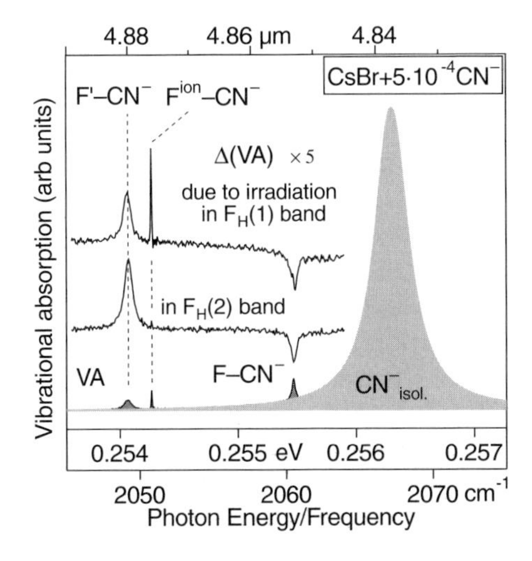

Fig. 6.2. Vibrational absorption of CN^--related centers in additively colored CsBr after F_H center aggregation measured at $T = 20$ K. The additional centers were produced by light irradiation at $T = 80$ K in the $F_H(1)$ and $F_H(2)$ bands. Difference spectra showing the effect of irradiation are shown on a scale expanded scale the ordinate

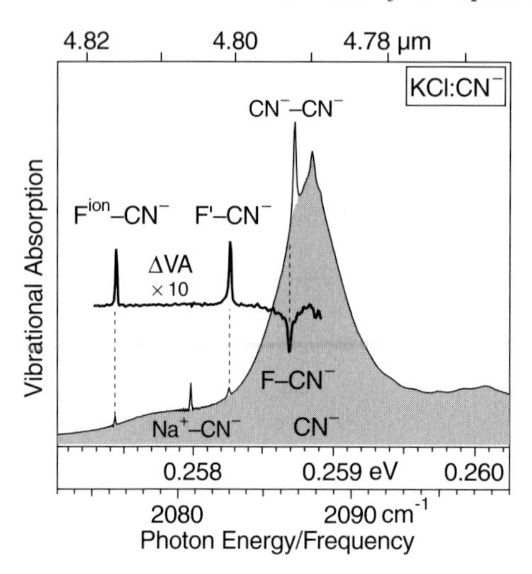

Fig. 6.3. Vibrational absorption of CN^--related centers in additively colored KCl after F_H center aggregation, measured at $T = 20$ K. The additional centers were produced by light irradiation at $T = 150$ K in the F_H band. A difference spectrum showing the effect of irradiation is shown on an expanded scale

Table 6.1. Fundamental eigenfrequencies $\overline{\omega}_{CN}$ and relative oscillator strengths of CN^- molecular defects with an anionic-vacancy neighbor occupied by 0, 1, or 2 electrons

Host	CN^- eigenfrequencies				oscillator strength ratio		
	CN^-_{isol} (cm^{-1})	F–CN^- (cm^{-1})	F'–CN^- (cm^{-1})	F^{ion}–CN^- (cm^{-1})	$\frac{f(F-CN^-)}{f(CN^-_{isol})}$	$\frac{f(F'-CN^-)}{f(CN^-_{isol})}$	$\frac{f(F^{ion}-CN^-)}{f(CN^-_{isol})}$
KCl	2087.8	2087.2	2083.0	2076.3	1	3	1
RbCl	2079.7	2079.1	–	2071.2	1	–	–
CsCl	2079.1	2068.8	–	2058.8	6	–	3
CsBr	2067.1	2060.5	2049.2	2050.8	5	10	3

6.3 Creation Kinetics

Using the simple picture illustrated in Fig. 6.1 the creation and destruction of F^{ion}–CN^- ("F_H^{ion}") and F'–CN^-("F'_H") centers under irradiation of light into the F_H absorption bands can be described by the following reaction equations:

$$F_H + h\nu_1 + F_H \rightarrow F_H^{ion} + F'_H\,, \tag{6.1}$$

$$F'_H + h\nu_2 + F_H^{ion} \rightarrow 2F_H\,. \tag{6.2}$$

If the photon energies $h\nu_1$ and $h\nu_2$ corresponding to the absorption peaks of the F_H and F'_H centers are clearly different, the reaction will go dominantly in one direction. Even if the absorption bands overlap, an equilibrium will be reached for which all three center types are present. From these equations, it is expected that both new center types will be created in equal numbers. However, this is only the case for KCl. For RbCl and CsCl, only the ionized defect type appears suggesting that the $F_H(CN^-)$ centers are not able to capture an electron and that other electron traps have to be present to accept the free electron. The number of the latter traps is limited, explaining the weak VA responses observed. In CsBr, the kinetics are even more complex, as they depend on the excitation energy. The observations can be explained if remnant F centers are included in the kinetics. The F band and the $F_H(2)$ band overlap strongly, such that both are photoionized. Three more reactions have to be considered in this case:

$$\mathrm{F} + h\nu_1 + \mathrm{F_H} \rightarrow \mathrm{F^{ion}} + \mathrm{F'_H}\,, \tag{6.3}$$

$$\mathrm{F_H} + h\nu_1 + \mathrm{F} \rightarrow \mathrm{F_H^{ion}} + \mathrm{F'}\,, \tag{6.4}$$

$$\mathrm{F} + h\nu_1 + \mathrm{F} \rightarrow \mathrm{F^{ion}} + \mathrm{F}\,. \tag{6.5}$$

These reactions show how F^{ion}–CN^- and F'–CN^- centers can be produced with the participation of F centers. This allows different relative numbers of the two defect types, depending on the excitation energy and aggregation stage. Moreover, the sum of the number of F^{ion}– CN^- and F'–CN^- centers produced is not necessarily equal to the number of F_H centers destroyed owing to the creation of F' centers, which are "invisible" in the VA. However, the creation kinetics under various experimental conditions can still be exploited to allow us to determine the relative numbers as well as the relative integrated absorption cross sections listed in Table 6.1. Let us now turn to the interpretation of the observed interaction effects.

6.4 Shift in Spectral Position

The spectral shift of the CN^- vibrational eigenfrequency in a crystalline environment compared with the free molecule is usually described in terms of a repulsive interaction with the surrounding ions (see Sect. 3.1.2). From that description one expects the eigenfrequency to be shifted to lower energies, compared with regular isolated CN^- defects, if one of the host ions is removed or replaced by a smaller one. From these considerations, we would expect that the spectral positions of all three centers (F^{ion}–CN^-, F–CN^-, and F'–CN^-) would be almost identical and shifted to the red relative to the isolated CN^- defect. The clear differences observed show that the size of the "available volume" cannot be the only source of the frequency shift. Additional effects have to be considered. A possible candidate is the electric field, which both F^{ion}–CN^- and F'–CN^- centers experience owing to the extra or missing electron on the neighboring anion vacancy. Such an energy shift has already been observed for CN^- molecules both in electric-dipole-ordered KCN and in measurements of the Stark effect [4, 5]. From these data, it can be extrapolated that a spectral shift of 10 cm^{-1} would require an electric field of 5×10^7 V/cm which is of the same magnitude as the field that can be estimated from a simple point charge model. Because the direction of the electric field created by an extra and a missing electron are opposite, the shifts of the two defects can only have the same direction if the relative molecular orientations are opposite. In a more quantitative treatment, one has to take into account the fact that both the CN^- and the neighboring host ions will adjust their position to minimize the overall energy of the system. This adjustment certainly will depend on the defect type and host material, explaining the observed differences in the relative spectral positions of the two defect types in KCl and CsBr.

6.5 Changes in Absorption Cross Section

As outlined in Sect. 3.1.2, the integrated absorption cross section can easily be changed in the vicinity of a polarizable defect. This model has been nicely confirmed for the complexes treated in this chapter. While the F^{ion}–CN^- experiences no enhancement effect compared to the isolated CN^- ion, the F–CN^- center shows an increase owing to the polarizability of the 1s and $2p_z$ states of the F electron. The effect is, however, most pronounced for the F'–CN^- center, for which the polarizability is even stronger owing to the more diffuse nature of the electronic states and the higher number of electrons. As we have seen in Chap. 4, any change of the transition probability will increase the E–V transfer efficiency. On the basis of the additional information obtained from our study of the F^{ion}–CN^- and F'–CN^- centers, we expect the most drastic change in the transition strength to occur for the F–CN^- center with the F electron in the RES. In this state, the electronic levels are very close to each other, allowing an effective mixing. A factor between 10 and 100 seems to be possible. Combined with a similar increase in the anharmonicity, the FD model may indeed be applicable to our E–V transfer problem, although its precondition of two independent defects is no longer valid for the mutual interaction parameters assumed. For this reason, the apparent agreement has to be treated with care.

References

1. M. Gregoriev, *F′ Centers in Alkali Halides*, of Lecture Notes in Physics, Vol. 298 (Springer, Berlin, Heidelberg, 1988).
2. W. Gellermann, Y. Yang, and F. Luty, Opti. Commun. **57**, 196 (1986).
3. W. Gellermann and F. Luty, Optics Comm. **72**, 214 (1987).
4. D. Durand, L. S. do Carmo, and F. Luty, Phys. Rev. B **39**, 6096 (1989).
5. R. Spitzer, A. Sievers, and R. Silsbee, J. Opt. Soc. Am. B **9**, 978 (1992).

7 $F_H(OH^-)$ Centers

Turning our attention from the F electron–CN^- complexes to those involving OH^-, the starting point will again be aggregated "$F_H(OH^-)$" centers, for which the two defects are located on neighboring lattice sites. While the E–V energy transfer cannot be observed directly for $F_H(OH^-)$ centers by the occurrence of vibrational luminescence (owing to the presence of effective nonradiative channels) and more difficult resonance Raman studies have to be employed, many of the other interaction effects are very pronounced for this defect type, and we shall put special emphasis on the following:

- Spectral shifts, both in the electronic and in the vibrational transitions.
- Optical and thermally driven bistability.

7.1 Cs Halides

Similarly to the situation for $F_H(CN^-)$ centers, there is a strong difference in the absorption behavior of $F_H(OH^-)$ centers between NaCl-type and CsCl-type host materials. Starting with the latter, the VA and EA absorption spectra (first studied by Krantz and Luty [1]) after center aggregation are shown in Fig. 7.1, in which drastic changes in the transition energies can be observed for both spectral regions.

7.1.1 Electronic Absorption

After center aggregation, the single F center absorption band is transformed into two absorption bands ($F_H(1)$ and $F_H(2)$), which are split by about 0.7 eV and are polarized dominantly parallel and perpendicular to the axis of the center, as indicated by the diagrams of the center in Fig. 7.1. The absorption spectra look very similar for the three Cs halides studied (CsCl, CsBr, and CsI). The spectral data are summarized in Tables 7.1 and 7.2.

Comparison of the shifts in the electronic absorption when the host is varied reveals that the splitting between the $F_H(1)$ and $F_H(2)$ bands is almost host-independent, despite its large value. This is quite unusual and puzzling. We shall encounter a similar behavior for the rare-earth-related RE^{2+}–CN^-complexes, where it is interpreted as a consequence of a charge

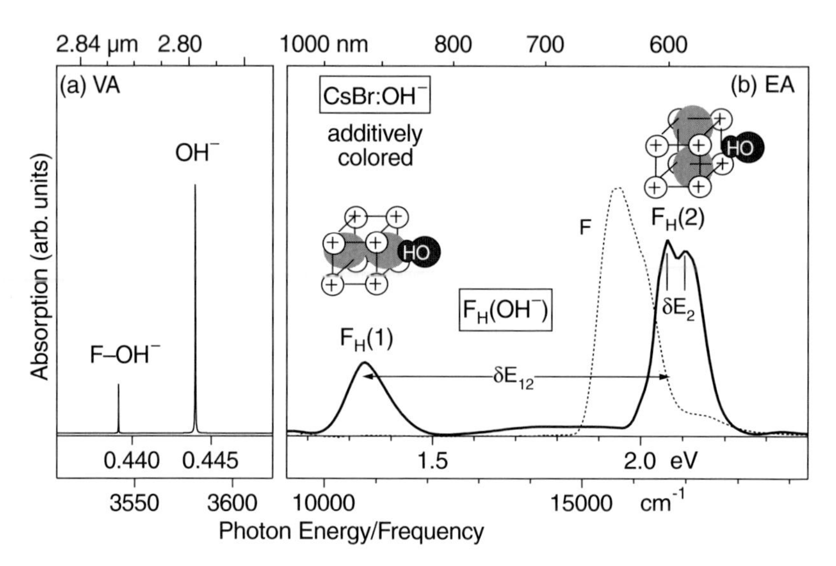

Fig. 7.1. $F_H(OH^-)$ centers in CsBr: (**a**) vibrational and (**b**) electronic absorption

Table 7.1. Electronic absorption energies and spin–orbit coupling parameters (Δ_F, Δ_2, Δ_{12}) of F, $F_H(OH^-)$ and $F_{H_2}(OH^-)$ centers in alkali halide hosts with CsCl structure. For the F_H center also the splittings (δE_2, δE_{12}) within the $F_H(2)$ band and between $F_H(2)$ and $F_H(1)$ are also given. The absorption energies have an error of ≈ 0.02 eV and the spin–orbit parameter of ≈ 6 meV

Host	F center		F_H center					
	E_0 (eV)	Δ_F (meV)	$F_H(2)$ (eV)	δE_2 (meV)	$F_H(1)$ (eV)	δE_{12} (eV)	Δ_2 (meV)	Δ_{12} (meV)
CsCl	2.14	−43	2.31	66	1.58	0.73	−48	−43
CsBr	1.96	−48	2.10	67	1.35	0.75	−51	−48
CsI	1.68	−55	1.84	61	1.1	0.73	−57	−53

redistribution among the defect partners. Such a charge redistribution is also observed in the laser-active $F_H(Tl^+)$, which is often also referred to as Tl^0 center [2, 3, 4, 5, 6]. Later on, when the ability of the OH^- to act as an electron trap is discussed (Chap. 8), it will become more evident that a similar charge transfer can occur within the $F_H(OH^-)$ center. Besides purely electrostatic effects, a charge redistribution will also cause changes due to electron–electron interaction, covalent effects, and exchange interaction. In a phenomenological picture, we can summarize these influences in terms of a crystal field, which is oriented along the axis of the center. The various level splittings of an F_H center predicted by a simple model are shown in Fig. 7.2.

Table 7.2. Vibrational transition energies, librational frequencies, and anharmonicity of $F_H(OH^-)$ centers in alkali halide hosts with CsCl structure. The data are compared with those for the isolated molecule. The differences are given as percentages

Host	Defect	OH⁻				OD⁻			
		$\overline{\omega}_{10}$	$\overline{\omega}_{20}$	$2\overline{\omega}_e x_e$	$\overline{\omega}_{libr}$	$\overline{\omega}_{10}$	$\overline{\omega}_{20}$	$2\overline{\omega}_e x_e$	$\overline{\omega}_{libr}$
CsCl	isol.	3602	7028	176	279	2655	5216	93	207
	F_H	3543.3	–	–	–	–	–	–	–
	$\frac{F_H - \text{isol}}{\text{isol}}$	−1.6%	–	–	–	–	–	–	–
CsBr	isol.	3580	6984	176	288	2642	5188	96	212
	F_H	3541.4	6889.9	192.9	367.3	2614.7	5127.5	101.9	266.6
	$\frac{F_H - \text{isol}}{\text{isol}}$	−1%	−1.3%	9.6%	27%	−1%	−1.2%	6%	26%
CsI	isol.	3572	6965	182	348	2636	5178	95	255
	F_H	3543.0	6898	188	414.8	2616	5133	99	303.2
	$\frac{F_H - \text{isol}}{\text{isol}}$	−0.8%	−1%	3%	19%	−0.8%	−0.9%	4%	19%

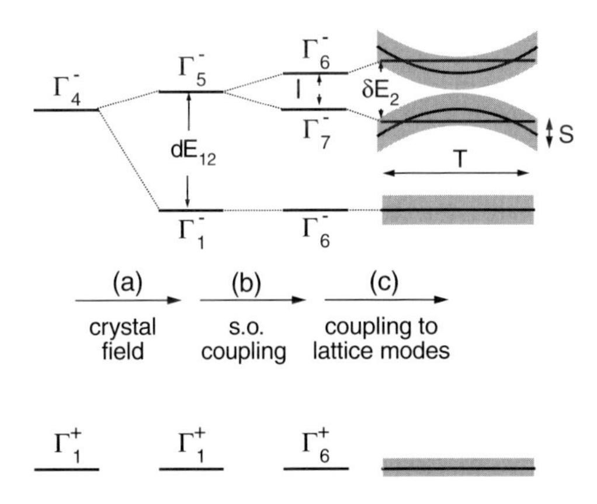

Fig. 7.2. Level diagram for a simple one-electron atomic model showing successively the excited-state splitting due to a tetragonal crystal field, spin–orbit (s.o.) coupling and lattice-mode interaction for an $F_H(OH^-)$ center in cesium halides. For clarity, the splittings are exaggerated

Starting with the degenerate 2p (Γ_4^-) states of an F center, the by far strongest perturbation is due to the crystal field, which causes a splitting into states perpendicular to the axis of the center p_x, and p_y (Γ_5^-) and a state parallel to the axis p_z (Γ_1^-). A subsequent "switching on" of the spin–orbit interaction produces two effects: splitting of the upper state Γ_5^- into Γ_6^- and Γ_7^- with a separation of $\lambda = \frac{2}{3}\Delta_2$, and a coupling of the state Γ_7^- with the state Γ_1^- with a strength Δ_{12}. Both spin–orbit parameters Δ_2, Δ_{12} have been measured (see Table 7.1) by magnetically induced circular dichroism of the absorption (MCDA) [7]. The measured value of Δ_2 is too small to account for the splitting of the $F_H(2)$ band by δE_2. A similar discrepancy has also been found for the regular F center and was resolved by Moran [8] who showed that the electron–phonon coupling in combination with the spin–

orbit interaction can significantly increase the shift by means of a dynamic Jahn–Teller effect. If we consider non symmetrical Γ_3 lattice modes (with coupling strength γ and coordinate T) and spin–orbit interaction for our $F_H(OH^-)$ center simultaneously in first-order perturbation theory, we find that the twofold-degenerate state Γ_5^- splits by $\sqrt{\lambda^2 + \gamma^2 T^2}$. Although the magnitude of this splitting varies with T it is always larger than λ and produces a time-averaged splitting of δE. Besides this repulsion effect, the coupled mode broadens the transition. The additional broadening by other modes is indicated by the shaded area in Fig. 7.2. As is common for crystal-field-based interpretations, the physical cause of the phenomenological crystal field remains obscure. Gash [9] and West et al. [10] tried to fill this gap by interpreting the shifts on the basis of electrostatic interactions. Gash considered a mixing of the 1s and $2p_z$ states by the permanent and displacement dipole moments of the OH^-. However, the results are unable to account for the ratio of the absorption strengths between the $F_H(1)$ and $F_H(2)$ bands. West et al. [10], on the other hand, considered mixing with the 2s state. Although these authors could fit the absorption spectra, the significance of their result seems doubtful for several reasons:

- The number of adjustable parameters is high.
- West et al. did not take the repulsion effect of non symmetrical modes described above into account.
- West et al. used spin–orbit parameters for the F, not the $F_H(OH^-)$, center.
- A similar model for the $F_H(CN^-)$ center has been shown to fail.
- Probably most importantly, the almost host-independent splitting of the $F_H(1)$ and $F_H(2)$ bands in Cs halides comes out purely accidental by non-systematic differences in the parameters.

A common feature of both theoretical treatments is the result that the permanent electric dipole (i.e. the H side) of the OH^- ion should point to the F center as indicated in Fig. 7.1.

Owing to the shortcomings of the theoretical models described above, it is fair to conclude that *the origin of the splitting of the* $F_H(OH^-)$ *absorption band is still not completely clear.*

7.1.2 Magnetic Resonance

In principle, further insight can be obtained by determination of the electronic charge distribution of the F electron by optically detected ENDOR. Unfortunately, no such investigations have been completely performed yet [11], so that only preliminary conclusions can be drawn. From the increased half-width of the optically detected EPR signal of the $F_H(OH^-)$ center compared to that of the regular F center (see Fig. 8.4) it can be concluded that the electron has some increased super-hyperfine (SHF) interaction with some of its Cs neighbors (which are essentially responsible for the width). In ENDOR spectra, Cs-related resonances at frequencies lower than for the F center are found, indicating a smaller SHF interaction. These two observations suggest a preliminary picture in which the electron is shifted to an off-center position within the anion vacancy, leading to an decreased interaction on one side and an increased interaction on the other, leading in turn to high-frequency ENDOR lines beyond the measured range [12].

7.1.3 Vibrational Properties

The vibrational properties of the centers are summarized in Table 7.2. In contrast to the EA results, a clear trend among the host materials can be found. With increasing anion size, the changes are reduced but are still significant, especially in the anharmonicity and the librational frequency. In agreement with the above assumption that the F electron is shifted within the anion vacancy, the vibrational changes for $F_H(OH^-)$ are in the same direction (but less pronounced) as those for $(OH^-)^-$ defects, in which the electron shares an anion vacancy with the OH^- molecule and which will be discussed in detail in Chap. 8. The $F_H(OH^-)$ center appears to be a case in between OH^- and $(OH^-)^-$, suggesting a shift of the electron distribution towards the molecule. The attractive force on the F electron could be caused by a fixed off-center position of the OH^- molecule. However, this intuitive picture still awaits confirmation.

7.1.4 The Relaxed Excited State

In view of this tendency to electron redistribution it is not surprising that the electronic luminescence and photoinduced ionization, both of which reflect the properties of the RES, are very complicated and depend strongly on the host material and the excitation conditions. Even more than for the $F_H(CN^-)$ center we have to assume that the RES is very different from that of a regular F center.

7.2 F_{H_2} Center

By means of further aggregation performed at $T = 180$ K with light irradiation in the $F_H(2)$ band, higher aggregates can be obtained for CsCl and CsBr, which reveal an even stronger splitting of the absorption band (Table 7.3). Using MCDA measurements and variation of OH^- concentration, these aggregates can be identified as an F center with two OH^- neighbors [13]. On the basis of experiments involving polarization-dependent bleaching and measurements of the resulting linear dichroism a $\langle 100 \rangle$ center arrangement can be found. This leaves two possibilities for the arrangement of the molecules: (1) on one side of the F center as $\langle 100 \rangle$ and $\langle 200 \rangle$ neighbors, or (2) as two

Table 7.3. Electronic absorption energies of $F_{H_2}(OH^-)$ centers in alkali halide hosts with CsCl structure, in eV

Host	$F_{H_2}(2)$ (eV)	$F_{H_2}(1)$ (eV)	Additional shift $F_{H_2}(1)$–$F_H(1)$
CsCl	2.33	1.31	0.27
CsBr	2.13	1.14	0.21

⟨100⟩ neighbors on opposite sides. A distinction between these options is very difficult especially if the orientation of the additional molecular is included as an extra degree of freedom. In ODESR measurements, the tendency for the width to increase is somewhat stronger than for the regular $F_H(OH^-)$ center, suggesting a further shift of the F center electron distribution. This observation favors option (1).

7.3 K and Rb Halides: Optical Bistability

7.3.1 Electronic Absorption

Similarly to the situation as for CN^- molecular defect partners, the spectral changes of the $F_H(OH^-)$ electronic transitions are less pronounced in the host materials with the NaCl structure. More interesting than these shifts is the observation that the centers occur in two distinctly different configurations,

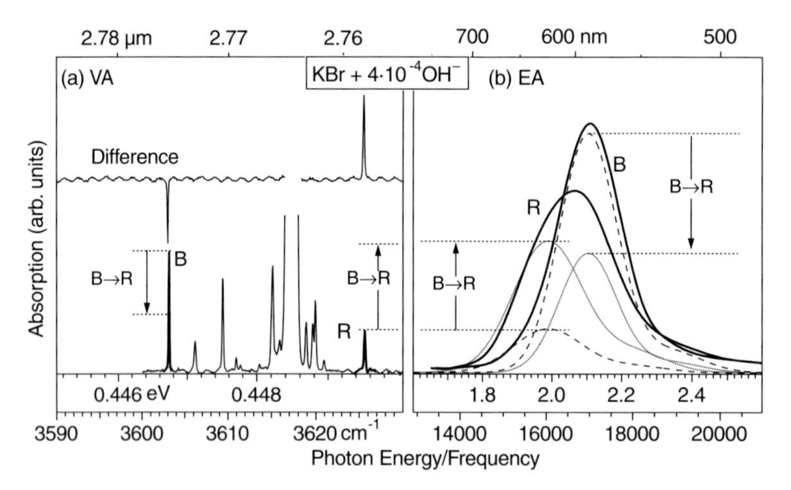

Fig. 7.3. (**a**) VA and (**b**) EA of $F_H(OH^-)$ centers in KBr. The changes induced by an optical B→R conversion at $T = 4$ K are indicated by arrows. On the basis on the VA results the EA bands before and after conversion have been decomposed into their B and R components (*dotted* and *dashed lines*)

which result in different EA spectra [14]. Compared with the regular F center, one center type referred to in the following as the blue center (B) absorbs at higher energies while the other type, referred to as the red center (R), absorbs at lower energies (see Fig. 7.3 and Table 7.4).

Table 7.4. Electronic-absorption energies in eV of F centers, and $F_H(OH^-)$ centers in the blue and red configurations in alkali halide hosts with NaCl structure. The values for KBr and RbBr were obtained by decomposition of the measured absorption spectra using the relative abundances of center types determined by VA measurements. HT describes the values for high temperature configuration. Details see text

Host	F center	F_H center		
		B	R	HT
KCl	2.3	–	–	2.2
KBr	2.06	2.11	2.0	–
RbBr	1.85	1.88	1.83	–

In KBr and some other halides, the two center types can be converted with high quantum efficiency (see Table 7.5) by light irradiation at low temperature. The resulting distribution of center types is only stable at very low temperatures; heating of the samples above $T = 30$ K and subsequent cooling down results again in a thermal-equilibrium distribution. Although these optical and thermal bistability phenomena were (accidentally) found and studied first in EA, a quantitative investigation is difficult because a clear separation of the two centers cannot easily be obtained owing to the broad F-type bands.

Table 7.5. Fundamental frequencies $\overline{\omega}$ of isolated $OH^-$and OD^- ions and and of the $F_H(OH^-)$ or $F_H(OD^-)$ center in the red ($\overline{\omega}_{\mathrm{red}}$) and blue ($\overline{\omega}_{\mathrm{blue}}$) configurations; all given in cm^{-1}, the ratio of vibrational oscillator strengths f_R/f_B, the concentration ratio in thermal equilibrium at $T = 4$ K $\frac{[R]}{[B]}$, and the conversion efficiencies $\eta_{R\to B}$ and $\eta_{B\to R}$

Host	isolated	$F\text{–}OH^-$					
	$\overline{\omega}$	$\overline{\omega}_{\mathrm{red}}$	$\overline{\omega}_{\mathrm{blue}}$	$\frac{f_R}{f_B}$	$\frac{[R]}{[B]}$	$\eta_{R\to B}$	$\eta_{B\to R}$
$KBr:OH^-$	3617.4	3625.6	3603.2	2±0.5	1:4	0.2±0.1	0.31±0.1
$KBr:OD^-$	2668.3	2674.4	2658.5	3.3±0.5	1:8	0.32±0.1	0.57±0.2
$RbCl:OH^-$	3633.2	3640.5	–	–	20:1	–	–
$RbBr:OH^-$	3610.3	3618	3595.9	2.9±0.5	10:1	0.21±0.1	0.51±0.2
$RbI:OH^-$	3595.3	3600.5	3586.7	3.5±0.5	10:1	–	–

7.3.2 Vibrational Absorption

A much better distinction can be achieved in VA, where the corresponding vibrational transitions of the two centers can be measured at well-separated spectral positions, shifted to lower energy (B) and higher energy (R) compared with the isolated molecular defect (see Fig. 7.3a and Table 7.5). A more detailed study of the bistability was possible on this basis. Let us concentrate first on the KBr host material. Similarly to the situation in EA, conversions of B to R and R to B can be observed; these result in difference spectra (shown in Fig. 7.3 for the B to R conversion), from which the ratio of the integrated VA cross sections can be determined. Using these values and assuming that they are constant, one can translate the measured VA by spectral integration into the relative numbers of centers N_R and N_B under various conditions. For instance, this technique makes it possible to separate (as shown for KBr in Fig. 7.3) the EA responses of the R and B centers by fitting, using the N_R and N_B values determined from the VA. Temperature dependence of the conversion process was studied, and it was found that the temperature scale can be divided into three ranges:

I. $T < 5$ K: in this range both center types are stable and each can be converted into the other type by light irradiation in the R or B absorption band, respectively.
II. $T = 5$–12 K: in this interesting range the relative numbers of centers change while their sum (N_B+N_R) stays constant. If we start from thermal equilibrium at $T = 4$ K, the R centers increase at the expense of the B-type centers. If an optical B to R conversion has been performed at $T = 4$ K, the relative numbers return with temperature-dependent rates to their thermal-equilibrium values. While at $T = 5$ K this process requires at least 30 min, at $T = 12$ K the thermalization is finished in less than a minute. This behavior is reflected in Fig. 7.4, which shows measurements taken at equal time intervals during heating up of the samples, by an increasing slope of the $N_B(T)$ and $N_R(T)$ curves.
III. $T > 12$ K: the sum of the relative numbers of R and B centers decreases.

VL is also a convenient tool to find other host materials for which the centers are bistable, in addition to the materials (KBr, KI) in which bistability is easily observed in EA [14]. In this way bistability was found in the RbBr and RbI hosts. Less successful was the search for bistability in RbCl, for which, despite many attempts of a R→B conversion at $T = 4$ K, only an R center could be found. In KCl, no VL line associated with an R or B center could be found at all. Comparing the behavior of the VL lines and also comparing the results of thermal/optical cycles shown in Fig. 7.4, we find many similarities among the various bistable systems [15]:

- In all three cases the changes of the VA line frequency $\Delta\bar{\omega}$ for the OH^- stretch band are in a direction opposite to the corresponding blue and red shifts of the electronic transition.

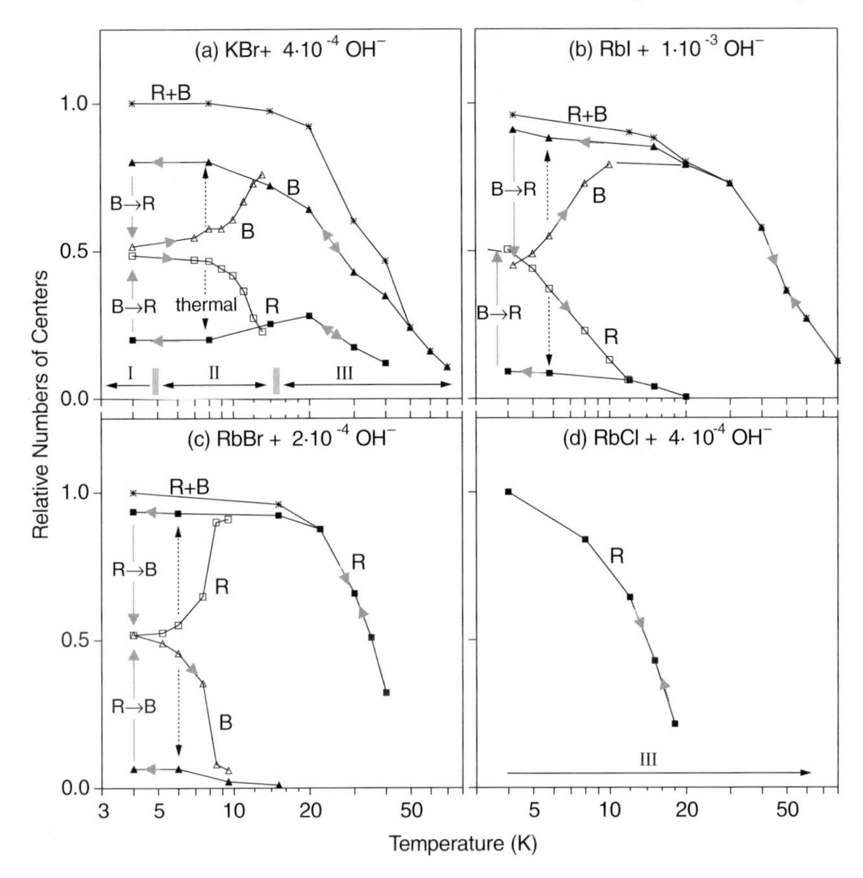

Fig. 7.4. Relative numbers N_{R} and N_{B} as a function of temperature (plotted logarithmically) in four hosts (**a**) KBr, (**b**) RbI, (**c**) RbBr and (**d**) RbCl measured in two ways: (i) reversible thermal cycle $4 \rightleftharpoons 80$ K (*solid symbols*) and (ii) interruption of the cycle at 4 K by an optical conversion, and subsequent heating in the dark (*open symbols*). The measurements were performed consecutively at constant time intervals of 5 min

- The R center shows a considerably higher oscillator strength.
- The VA line of the R center broadens with temperature considerably more than the B center line.
- The sum of the number of the two center types remains constant for temperatures in ranges I and II and decreases with increasing T in range III. Both types of centers disappear completely at higher temperature ($T = 50$–$80\,\mathrm{K}$) but return after cooling down. This property is also shared by the $\mathrm{VA_R}$ response in RbCl.

Only the details of the preferences of the various hosts for the R and B configurations are different: in KBr and RbI, the B type center completely dominates the thermal equilibrium at $T = 4$ K (see the ratios in Table 7.5), while in RbBr the R center is the dominant type.

7.3.3 Microscopic Structure

The microscopic structure of the bistable F–OH^- pairs in KBr has been studied by magnetic resonance [16, 17, 18, 19]. In these ENDOR/ODENDOR experiments[1], several new resonance lines appeared or changed as a result of optical F→F_H conversion and subsequent B→R conversion. They could be assigned to the hydrogen nuclei of the OH^- ions and to the potassium neighbors. The oxygen could not be detected owing to the nearly complete absence of the magnetic ^{17}O isotope. Analysis of the values, signs, and symmetries of the measurable SHF-interaction tensor yielded – besides the general (200) site symmetry of the pair – more detailed information about the hydrogen nuclei and the nearest-neighbor (NN) K^+_γ ion located on the pair axis between the F center and the OH^-defect. These microscopic details are illustrated in Fig. 7.5a.

- In the R configuration, the distances of the K^+_γ and H nuclei from the F electron are found to be 3.8 and 7.4 Å, respectively, which are considerably increased compared with the normal lattice position. The observed positive sign of both the isotropic and the anisotropic hydrogen SHF interaction indicates alignment of the OH^- ion parallel to the defect axis with the H nucleus pointing away from the F center.

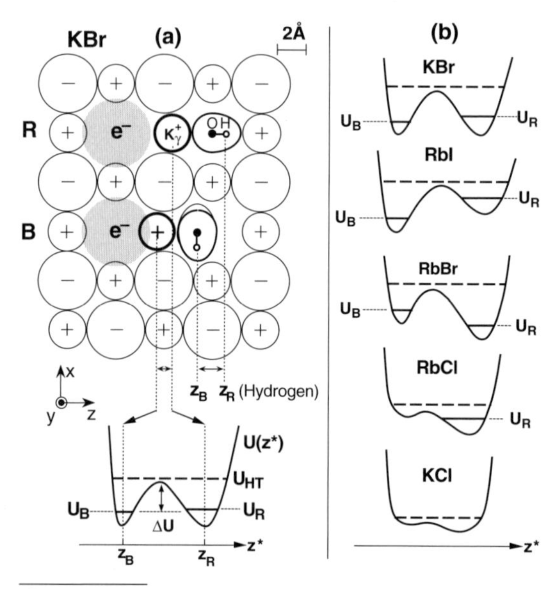

Fig. 7.5. (**a**) view of a (100) plane in KBr showing in the size and positions of the three constituents of the $F_H(OH^-)$ center in its R and B configurations. (**b**) Schematic bistable potentials in various host materials

[1] For an overview of the various techniques of magnetic resonance and a review of these data, see [19].

- In the B configuration, the distances of the K_γ^+ and the H nuclei from the F electron are found to be 2.8 and 5.4 Å, respectively, considerably smaller than for the regular lattice sites. In contrast to the R configuration, a negative sign of the isotropic SHF parameter of hydrogen is found, understandable so far only on the basis of exchange polarization effects [20]. This requires an OH^- orientation perpendicular to the $\langle 200 \rangle$ center axis. As the lower symmetry of this arrangement is not observed in ENDOR, a rapid ($\tau^{-1} = 10^6$ s^{-1}) center reorientation within the xy plane perpendicular to the center axis is assumed.

This model has been challenged in terms of the precise OH^- orientation by recent measurements of the polarization behavior of the VA lines under application of uniaxial stress [21]. In these experiments, a nonthermal distribution of R and B centers is prepared by optical means (without stress) at very low temperature and the sample is then heated under uniaxial stress into the temperature range II in which the thermal equilibrium is reached slowly. Because of the stress, the relaxation rate is anisotropic such that centers with different orientations relative to the $\langle 100 \rangle$ axis of the stress relax at different rates. By means of this trick, the equality of the [100], [010], and [001] directions is removed. In the resulting change of the VA spectra a $\langle 100 \rangle$ dichroism is observed, which has the same sign for both center types, showing that the OH^- molecules in the R and B configurations are oriented dominantly parallel to each other. A more detailed evaluation taking account of the different vibrational oscillator strengths and the measured degree of polarization reveals that (in a time average) the relative angle between the two arrangements must be less than 45°. Possibly this is a result of a more complicated reorientational movement within the R or B configuration.

Despite this uncertainty in the OH^- orientation, the positions of the K_γ^+ ion are accurate and can be used to qualitatively understand the relative energy shifts for the R and B configurations. For both EA and VA, empirical rules exist which relate the transition energies to the average distance d to the respective neighbors of the F center and the OH^- ion. While for the F electron absorption this rule can be expressed by the well-known Mollwo–Ivey relation [22] $E \propto d^{-2}$, for the OH^- vibrational absorption it is well established from uniaxial stress experiments [23] and host material variation and by theory [24] that for example a reduction of d leads (owing to increasing repulsion) to a higher VA transition energy. This trend is quantified in the empirical relation (3.6) in Sect. 3.1.2. The observed shifts in the EA and VA can be accounted for in this way because the off-center displacements of the K_γ^+ ion have opposite effects on the F electron and the OH^- molecule. Beyond these very plausible general trends in the EA and VA frequency shifts, however, it is much harder to answer more detailed questions, for instance "Why does the very large inward or outward shift of the K_γ^+ not split the EA transitions for polarizations parallel and perpendicular to the center axis?" Other than a small asymmetry of the band, no indication of this is observed.

In his pseudopotential-based theoretical treatment and interpretation of the EA and magnetic-resonance results, Gash [25] considered, within a point ion model, the splitting of the excited state into two and three nondegenerate components for the red and blue center types, respectively, and could show that the corresponding subbands overlap strongly such that they cannot be resolved owing to their considerable homogeneous width.

7.3.4 Entropy-Driven Bistability: Two-Center Model

On the basis of the EA and magnetic-resonance results, a two-center model was proposed in which the center configuration at high T was assumed to be the same as the R configuration. In order to explain the bistable behavior, the Gibbs free energy $G = H - TS$ has to be considered which includes, in addition to the configurational enthalpy H, also the entropy $S = k_B \ln(W)$ where W represents the number of microscopic states. At $T = 0$ K the configurational enthalpy H determines the free energy, and we assume that in KBr this is lower for the B configuration. As the temperature increases, the entropy term becomes important and may invert the relative position of the free energy if the entropy of the R configuration is bigger than that for the B. While for the latter the number of equivalent configurations is given by the four OH^- orientations relative to the center axis, the entropy factor for R is determined by the temperature-dependent soft vibrations of the K_γ^+ and the OH^- molecule, and may become bigger as T is increased.

In the VA measurements, evidence for such an entropy-driven bistability behavior can be found in the temperature ranges II and III up to $T = 40$ K. For higher T, however, the total number of R and B centers reduces so strongly that this reduction can only be explained by the occurrence of a third type of center invisible in VA measurements. A phenomenological model that assumes three center types and their interconversion will be discussed in following.

7.3.5 Three-Center-Type Model

From the results of the ENDOR measurements, it became apparent that the $F_H(OH^-)$ center cannot be treated as a simple F-center–OH^- pair but that at least the potassium ion K_γ^+ in between has to be taken into account. This ion undergoes a large change in its position z during center conversion. However, this z motion of the K_γ^+ ion cannot be considered individually, but instead the location and orientation of the OH^- defect and of the F electron have to be included. The total energy of the system is then a sum of all contributions. To simplify the argument and to allow us to use a schematic picture like that in Fig. 7.5, it is still advantageous to consider only the single coordinate z but one needs to keep in mind that it is related to the other system coordinates. The effective potentials $U(z^*)$ determined in this way can be thought of as properly chosen cross sections through the multidimensional coordinate space. The stable configurations R and B correspond to the minima of a two-well potential. Owing to this more general meaning of the displacement coordinate, we represent it by the symbol z^*. Unlike the symmetric potentials of some well-known off-center defects (such as Li^+ in KCl), the off-center potential for our K ion with its two totally different neighbors F and OH is always asymmetric in terms of the depth and width of the two local minima, as indicated in Fig. 7.5. The lowest states, denoted by U_R and U_B, in these two wells must be considered as two different coupled K_γ^+–OH^- pairs, which can change their positions between the U_R and U_B minima only by thermally activated or optically induced motion over the barrier with a gradual change of the character of the coupling from the B to thc R configuration. Besides the ground states U_R and U_B, excited states due to K_γ^+–OH^- motions can exist which will be different in their nature and excitation energy within the two wells.

It is evident from the above discussion that the potential $U(z^*)$ is very sensitive to any changes of the constituents of the system and of the lattice constituents and spacings. Owing to the absence of any theoretical treatment that could be used for interpretation of the experimental results on the basis of calculated $U(z^*)$, we use the measured data to approximately predict the shape of the potential. This is illustrated in Fig. 7.5 for four host materials. The differences in the low-temperature B/R ratios are reflected in the depths of the potentials. While for RbI the B configuration is lower in energy, the R configuration has the deeper minimum in RbBr. If, additionally, the statistical weights of the center configurations are taken into account, the 4:1 ratio between the numbers of B and R found in KBr suggests that the wells are of almost equal depth in this host. The broader VA lines for the R-type centers are interpreted within this model as a coupling to soft modes, and hence we depict a wider potential well for this configuration compared with the more strongly orientationally aligned B-type. If T is increased into the temperature range II the thermal energy obviously becomes high enough to allow thermally activated reorientation over the energy barrier, but the two configurations still have to be considered as independent. With a further rise in temperature, a decrease of the number of B and R is observed, which can only be explained by the appearance of a new center configuration. In terms of our $U(z^*)$ model we assume that high temperature excited states U_{HT} (indicated schematically by dashed lines in Fig. 7.5) lying above the energy barrier are thermally occupied with increasing probability. In these states the K_γ^+ or Rb_γ^+ ion is less localized but becomes freer to perform linear vibrations over the whole available z range, thereby becoming similar to the regular K_γ^+ or Rb_γ^+ ion. As a consequence of such motion, the OH^- undergoes drastic changes in both its axial and its rotational behavior.

The new average K_γ^+ position yields a situation for OH^- which resembles that of OH^- as an isolated defect. Consequently, similar rotational behavior and transition energies are expected. For these reasons, the VA response of the system in this state will no longer consist of two distinct, rather sharp VA lines but will be hidden because of its rotational broadening under the unavoidable and much stronger absorption band of isolated OH^-, making it invisible in the VA experiments. We can also interpret the negative VA result for KCl in this way. Obviously, for this host, with its much smaller anionic constituents and smaller size misfit for the OH^- defect, the two-well potential is not deep enough for a localized ground state to exist allowing a rapid reorientation of the defect complex. All of these assumptions about the high-temperature structure of the $F_H(OH^-)$ center are supported by ENDOR measurements (for KCl), in which a z-motion of the K^+_γ ion with a large amplitude (which increases with temperature) and a time-averaged K^+_γ displacement away from the F center were found [26]. For RbCl, we assume a potential in which only the R configuration has a stable ground state, while the well for B is too shallow. If we compare the situations for Rb halides with different anions, we can note that potential well for the B configuration becomes successively deeper in the sequnence Cl $\rightarrow$ Br $\rightarrow$ I. Obviously, the off-center arrangement of the OH^- ion present in the B configuration becomes more likely as the anion size increases, just as for the isolated molecule On the basis of our model the generally observed redshift of the EA band for F_H compared with the F center can be explained, not as has been done frequently, as the R low-temperature configuration, but as the high-temperature configuration in which, owing to the small size of OH^-, the K^+_γ and Rb^+_γ neighbors are shifted *on average* towards the OH^- ion giving more space for the F electron and resulting in a lower transition energy.

The important role of K^+_γ and Rb^+_γ becomes even more apparent if the Cs halides are considered. In these hosts the OH^- ion is located at a $\langle 100 \rangle$ position with respect to the F electron without a cation in between, and consequently not a single indication of bistability has been observed.

7.3.6 Changes in Vibrational Absorption Cross-Section

The changes in absorption strength between the R and B configurations ($f_R/f_B > 2$) are at first sight contradictory to the model developed in Sect. 3.2.2 because of the smaller distance between the F center and the OH^- ion in the B center. In this simplified treatment, however, the orientations of the center axis and the molecule were assumed to be parallel, as in the case of the R configuration, giving a maximum electric-field change at the F center location. In the B configuration, the relative orientation is less favorable, reducing the influence of the dipolar electric field. Furthermore, the contributions from K^+_γ are expected to be higher in the case of the R center. The vibration of the OH^-molecule will also cause a z motion of K^+_γ and a redistribution of the F electron, all causing additional "oscillators" with the same frequency as the molecule and an increase in the absorption cross section. This interrelation will be most pronounced for a wide potential, as assumed for the R center.

7.4 E–V Energy Transfer

Owing to the dominance of nonradiative channels, for the vibrational excitation no vibrational luminescence has been observed for $F_H(OH^-)$ centers, with the exception of CsI. The presence of a still rather efficient E–V energy transfer for the $F_H(OH^-)$ center has been shown instead by anti-Stokes resonance Raman spectroscopy, in which transitions from vibrationally excited molecules could be detected. The most detailed measurements have been performed by Gustin et al. [27, 28] for KBr. It was shown through power variation that the E–V transfer occurs predominantly into $v = 1$ with a small fraction $\sim$ 10% into $v = 2$ for OD^-. Similar results are found for other alkali halide hosts with the fcc crystal structure (RbCl and KCl). Two interesting observations could be made in the Raman spectra which were not seen in VA:

- Vibrational excitations and the corresponding Raman transitions could be seen not only for the B and R configurations but also for a spectral position coinciding with that of the isolated defect. This can be interpreted as E–V transfer to not direct neighbors of the F center. We shall come back to this phenomenon in Sect. 8 treating it there for Cs halides where it is much more apparent.
- In KCl, a Raman line that most likely corresponds to the R configuration is observed, and in RbCl, evidence for the B configuration in addition to R could be found. This suggests that very weakly bound states exist for these "extra" configurations, which can only be populated under constant light irradiation, in competition with thermal reorientation. Full clarification of this difference between VA and Raman has not been achieved yet.

Unlike the situation for the three host materials KBr, RbCl, and KCl, mentioned above in which the Raman signal increases with center aggregation, the anti-Stokes Raman signal is reduced after F_H center aggregation in CsBr (see Chap. 8). This is consistent with the recurrence of F-type electronic luminescence [29]. The relative E–V transfer rates, however, behave quite similarly, as shown in the Raman spectrum presented in Fig. 8.9. Again the E–V transfer occurs predominantly into $v = 1$.

7.5 Dynamic Properties

The excitation/relaxation cycle of the $F_H(OH^-)$ centers in the alkali halides with the fcc crystal structure is even more complicated than for $F_H(CN^-)$ because it includes, besides configurational relaxation of the lattice and E–V transfer also the reorientation of the molecule.The time dependence of the excitation/relaxation process has been studied by several groups [30, 31, 32, 33, 34, 28]. Among those studies the comprehensive work of Gustin et al. [35] appears to be the most reliable, as they took special care to avoid multiple excitation of the same defect complex by using a low laser power in connection with a lock-in technique. They found that the transfer occurs within a few picoseconds, suggesting a transfer time which is faster than the time that the electron usually requires to reach the relaxed excited state. From this result, they concluded that the transfer occurs via a crossover *during* the relaxation into the RES. This conclusion can be illustrated in a configurational-coordinate diagram (Fig. 7.6). The difference in the eigenfrequencies of two OH^- and OD^- molecules causes a difference in the position of the crossover points, explaining the observed isotope effect on the relative transfer efficiencies. The E–V transfer during the crossover must be so efficient that essentially no electron reaches the RES and no transfer to $v = 3, 4\ldots$ is observed. It should be pointed out that the CC diagram for $F_H(OH)^-$ in KBr is a very rough approximation – even more than in other cases – because not only the 2p/2s mixing occurring during lattice relaxation but also the possible reconfiguration between R- and B-type centers have been neglected. However, these effects may become most important only on longer timescales so that the above picture should be valid at least in the very early stages, on timescales of the order of a picosecond, of the configurational relaxation.

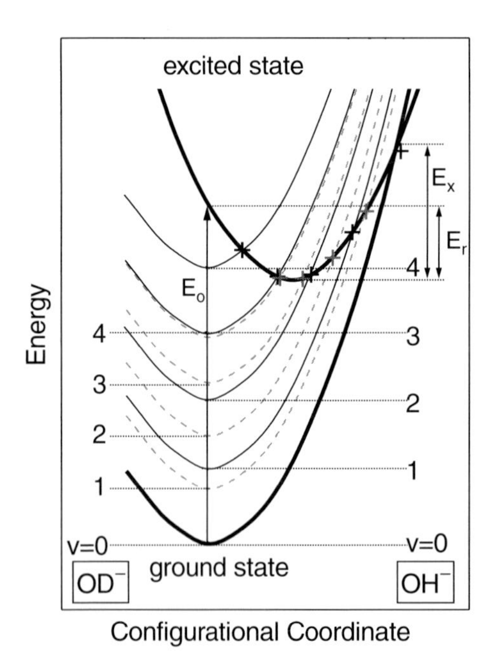

Fig. 7.6. Schematic CC diagram of the $F_H(OH)^-$ and $F_H(OD^-)$ centers in KBr. In addition to the potential curves of the ground and excited state (*thick lines*), vibrational excitations of OH^- and OD^- are indicated by additional parabolas (*thin lines* and *thin dashed lines*). The crossing points are indicated by *circles* (black, OH^-; gray, OD^-)

Besides the fast time component slower components (100–1500 ps) were observed in the ground state recovery measurements. These components are connected with the reorientation of the defect, which causes the optical bistability and the nonradiative vibrational relaxation of the molecule. It is generally assumed that the reorientation is correlated with the excitation of librational–rotational modes. Two stages of the excitation/relaxation process at which reorientation may occur have to be considered:

- during relaxation of the molecule, or
- during electronic relaxation before the relaxation of the molecule.

While the timedependent measurements did not span a wide enough range in time to answer this question, we favor the second option because the electronic relaxation is connected with a shift of the K_{γ}^{+} or Rb_{γ}^{+} cation located between the F center and the OH^{-}/OD^{-} molecule, which may play the role of a "gate" between the two configurations, as described in Sect. 7.3. Indeed, Gustin et al. observed a reorientation component within a correspondingly fast timescale. However, they could not determine the efficiency.

Although Gustin et al. observed a change of the absorption due to vibrational excitation, they could not deduce the coupling constants between the OH^{-} molecule and the F electron from their data because the measured ranges of both time and probe energy were too small to perform an evaluation similar to that of Samiec [36]. The lack of this crucial information makes it impossible to reliably check if the high transfer efficiencies can be explained within the crossover model.

References

1. M. Krantz and F. Luty, Phys. Rev. B **37**, 8412 (1988).
2. W. Gellermann, F. Luty, and C. Pollock, Opt. Commun. **39**, 391 (1981).
3. E. Goovaerts, E. Andriessen, S. Nistor, and D. Schoemaker, Phys. Rev. B **24**, 29 (1981).
4. P. Baranov and V. Khramtsov, Phys. Status Solidi (B) **101**, 153 (1980).
5. F. Ahlers, F. Lohse, J.-M. Spaeth, and L. F. Mollenauer, Phys. Rev. B **28**, 1249 (1983).
6. W. Joosen, E. Goovaerts, and D. Schoemaker, Phys. Rev. B **32**, 6748 (1985).
7. V. Dierolf, H. Paus, and F. Luty, Phys. Rev. B **43**, 9879 (1991).
8. P. Moran, Phys. Rev. **137**, A1016 (1965).
9. P. W. Gash, Phys. Rev. B **35**, 774 (1987).
10. J. West, K. T. Tsen, and S. H. Lin, Mod. Phys. Lett. B **9**, 1759 (1995).
11. V. Dierolf and J.-M. Spaeth, Mat. Science Forum (Proc. ICDIM96) **239–241**, 461 (1997).
12. V. Dierolf, T. Pawlik, and J.-M. Spaeth (unpublished).
13. M. Krantz, F. Luty, V. Dierolf, and H. Paus, Phys. Rev. B **43**, 9888 (1991).
14. G. Baldacchini, S. Botti, U. M. Grassano, L. Gomes, and F. Luty, Europhys. Lett. **9**, 735 (1989).

15. V. Dierolf and F. Luty, Phys. Rev. B **54**, 6952 (1996).
16. H. Söthe, J.-M. Spaeth, and F. Luty, Rev. Solid State Sc. **4**, 440 (1990).
17. H. Söthe, J.-M. Spaeth, and F. Luty, Radiat. Eff. Defects Solids **119–121**, 269 (1991).
18. H. Söthe, J.-M. Spaeth, and F. Luty, J. Phys.: Condens. Matter **5**, 1957 (1993).
19. J.-M. Spaeth, J. Niklas, and R. Bartram, *Structural Analysis of Point Defects in Solids*, Springer Series: Solid-State Sciences 43 (Springer-Verlag, Berlin, Heidelberg, New York, 1992).
20. F. J.Adrian, A. N. Jette, and J.-M. Spaeth, Phys. Rev. B **31**, 3923 (1985).
21. F. Luty and V. Dierolf (unpublished).
22. H. Ivey, Phys. Rev. **72**, 341 (1947).
23. H. Härtel, Phys. Stat. Sol. **42**, 369 (1970).
24. G. Pandey and D. Shukla, Phys. Rev. B **4**, 4598 (1985).
25. P. W. Gash, Material Science Forum **239-241**, 373 (1997).
26. M. Jordan, H. Söthe, J.-M. Spaeth, and F. Luty, in *Proceedings of the International Conference on Defects in Insulating Crystals* (Parma, Italy, 1988), p. 11.
27. E. Gustin, M. Leblans, A. Bouwen, and D. Schoemaker, Phys. Rev. B **54**, 6963 (1996).
28. E. Gustin, Ph.D. thesis, Universitaire Instelling Antwerpen (UIA), 1995.
29. M. Krantz, Ph.D. thesis, University of Utah, 1987.
30. D. Jang, T. Corcoran, M. El-Sayed, L. Gomes, and F. Luty, in *Ultrafast Phenomena V*, edited by G. Fleming (Springer, Berlin, 1986), p. 208.
31. D.-J. Jang and J. Lee, Solid State Commun. **94**, 539 (1995).
32. M. Leblans, E. Gustin, A. Bouwen, and D. Schoemaker, J. Lumin. **58**, 388 (1994).
33. M. Casalboni, P. Prosposito, and U. Grassamo, Solid State Commun. **87**, 305 (1993).
34. E. Gustin, W. Wenseleers, M. Leblans, A. Bouwen, and D. Schoemaker, Radiat. Eff. and Defects Solids **134**, 489 (1995).
35. E. Gustin, M. Leblans, A. Bouwen, and D. Schoemaker, Phys. Rev. B **54**, 6977 (1996).
36. D. Samiec, Ph.D. thesis, Universität GH Paderborn, 1997.

8 Interaction Between F Electrons and Distant OH^- Molecules

In this chapter,[1] we deal with the effect of long-range interaction between an F electron and an OH^-. The strength of the effect is quite surprising, and suggests that the behavior of an OH^- molecular defect shows many similarities to that of an F center. From this viewpoint the quenching of F luminescence in OH^--doped samples can be readily explained by analogy to the well-known quenching of the emission at high F-concentrations.

8.1 The Main Idea

The first electronic–molecular defect system studied was the F center in combination with an OH^- ion because, owing to its high permanent electric dipole, it seemed most promising for achievement of the initial goal (i.e. to optimize the F center luminescence behavior; see Chap. 1). However, the expectations were not fulfilled. On the contrary above a certain OH^- concentration the F luminescence was quenched completely, even before any center aggregation was performed and even at the lowest temperatures [1]. When the OH^- concentration was increased an enhanced efficiency for photoionization and a higher photoconductivity were observed, along with a reduction in the lifetime and quenching of the F center luminescence. This behavior resembles the observations in strongly additively colored samples with F center concentrations above 10^{18} centers/cm^3. In this case, the behavior could be explained consistently by the formation of F′ centers through electron tunneling [2]; this was demonstrated most convincingly by a "switch-off" of the tunneling and reappearance of the F luminescence caused by the spin polarization achieved in a magnetic field [3]. Guided by these similarities, Luty postulated that the OH^- molecule could play a role similar to that of an F center and might be able to function as an electron trap. In order to visualize these ideas and the underlying processes, we use here a schematic energy diagram (Fig. 8.1), in which the energy of the F electron (and not the total energy) is drawn relative to the conduction band. After optical excitation by light irradiation in the F band, the F electron rapidly relaxes into a diffuse RES which lies

[1] Most of the work reported in this chapter is the product of a collaboration with Professor F. Luty at University of Utah.

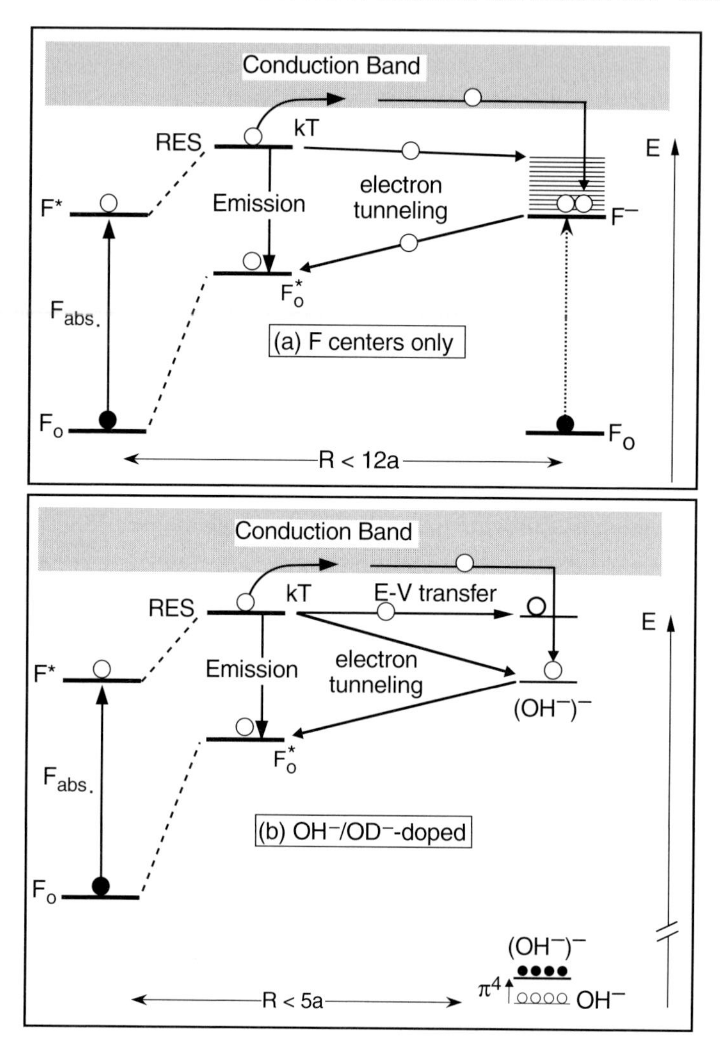

Fig. 8.1. Schematic model for the formation of F′ and $(OH^-)^-$centers through thermal activation or electron tunneling. In the case of $(OH^-)^-$ the possibility of an E–V energy transfer is indicated as well. See text fro details

fairly close to the conduction band. At elevated temperatures, a thermally activated ionization can occur, producing a free electron which has a chance to be captured by another F center, producing an F′ center. At higher F center concentrations, an F′ center can be produced temporarily even at very low temperatures through electron tunneling. A fast back-tunneling then shortcuts the F emission, explaining the luminescence quenching. The extra energy is distributed into phonons. If the second, electron-accepting, F center is now replaced by an OH^- molecule, not only phonons but also the local modes of the molecule are available, which could lead to vibrational excitation. This type of E–V transfer would be possible during both the forward-tunneling and the back-tunneling. The argument up to this point is just a speculation based on some analogies in the behavior of F centers and OH^- molecules with respect to their ability to quench F center luminescence. However, a test to show that processes of this kind and an E–V energy transfer really occur has been performed in several steps:

- Find, identify, and characterize OH^-/OD^-defects which have captured an extra electron through thermally activated photoionization of F centers.
- Show that in highly OH^--doped samples, these $(OH^-)^-$ defects can also be produced at very low temperatures.
- Show that E–V transfer occurs even for nonaggregated (statistically distributed) defects.

The extensive studies leading to the proof of all these points are summarized in the following.

8.2 OH^- Defects with a Captured Extra Electron

The idea of trapping an electron at substitutional defect sites in alkali halides is not new. For instance, it was found that cationic defects (e.g. Na^+ in KCl) are able to weakly bind electrons [4, 5]. These defects could only be investigated in their electronic properties such that an identification was very difficult.

8.2.1 Absorption Results

In contrast to those defects the electron trapping by our molecular defects should become apparent in several spectral regions:

- *In the UV*, by the appearance of the α band owing to an anion vacancy and a change of the OH^--related electronic absorption.
- *In the visible or the NIR*, owing to the loosely bound electron. Additionally, the F band should be reduced.
- *In the MIR*, owing to the vibrational absorption of the perturbed OH^- molecule.

By comparing these spectral responses, a correlation between the electron donor and acceptor can be achieved.

Electronic Absorption. The effects on the absorption in the UV and VIS are illustrated for the OH^- molecule in additively colored CsBr in Fig. 8.2. Thermally activated photoionization of the F band is possible in this host material at temperatures above $\sim$ 80 K. In comparison with the absorption of the F′ centers produced in samples which have not been doped with OH^-, the shape and position of the absorption spectra connected with the electrons trapped by OH^- related defects are quite different. In common with F′, on the other hand, the absorption can easily be bleached by light irradiation, even at very low temperatures and the centers are unstable at higher temperatures (T = 140 K for one type and 170 K for another type) even in the dark. Using these properties, at least two new center types can be distinguished; they will later on be assigned, as indicated in the Fig. 8.2, to electrons captured in the vicinity of one or two OH^- molecules ($(OH^-)^-$ and $(OH^-)_2^-$).

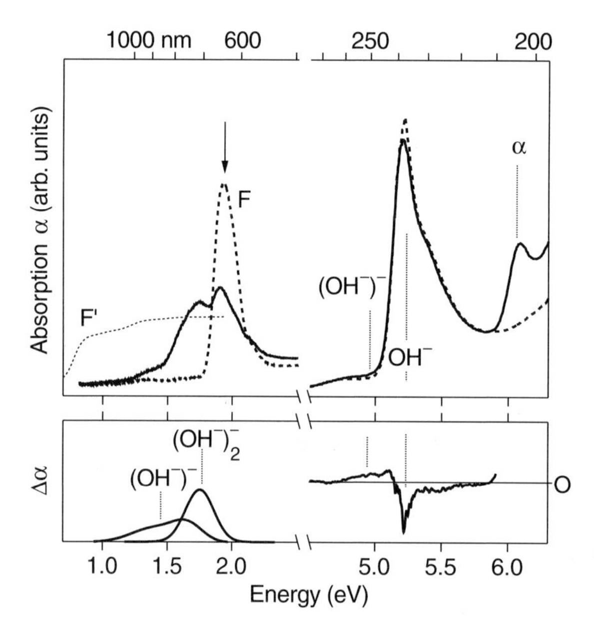

Fig. 8.2. *Top*: Electronic-absorption spectra before and after irradiation with light in the F band at T = 80 K. (The F′ band, which is found under the same experimental conditions in samples without OH^-, is shown for comparison).
Bottom: Decomposition into absorption contributions from $(OH^-)^-$ and $(OH^-)_2^-$, and difference spectra for the UV region

The shape of these absorption bands can be evaluated by a method devised by Malghani and Smith [6, 7], which links a combination of certain moments of the absorption spectrum $K(\omega)$ with the mean square value r_{rms} of the radial electron distribution through the Vinti sum rule [8] as follows

$$r_{\mathrm{rms}}^2 = \langle 0| r^2 |0\rangle = \frac{3\hbar}{2m} \frac{\mu_{-1}}{\mu_0} ,$$

where

$$\mu_{-1} = \int_0^\infty \omega^{-1} K(\omega)\, \mathrm{d}\omega ,$$

$$\mu_0 = \int_0^\infty K(\omega)\,\mathrm{d}\omega\,.$$

In this equation m represents the effective mass of the electron considered. We find a value of $r_{\mathrm{rms}} \approx 2.8$ Å for the $(OH^-)^-$ center, which corresponds to a 20% increase relative to the regular F center, indicating a less localized electronic distribution.

The absorption of OH^- in the UV also shows minor differences, with a tendency for the perturbed center to absorb at lower energies. Since the number of electrons available is limited by the low F center concentration and is much lower than the number of OH^- molecules, the dominant UV absorption is still due to the unperturbed OH^- molecules. Some of these EA responses were observed in earlier studies [9, 10, 11] on this type of sample but could not be identified at the time. As we have seen before for other systems, the assignment can be done most reliably with the VA of the OH^--related centers.

Vibrational Absorption. As shown in Fig. 8.3a, two new VA peaks appear after laser irradiation in the F band at $T = 80$ K. The most apparent features

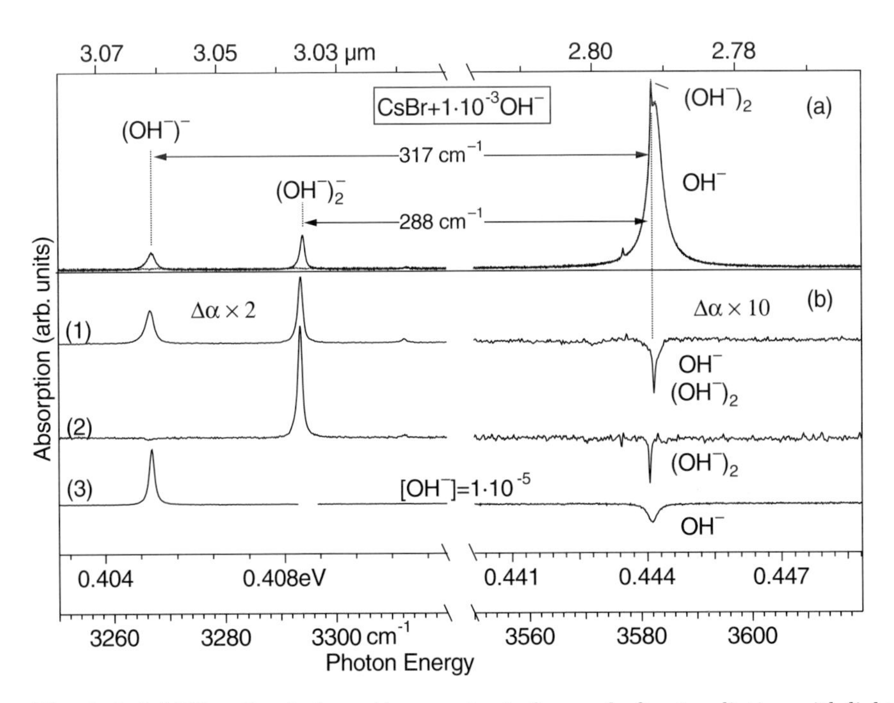

Fig. 8.3. (**a**) Vibrational-absorption spectra before and after irradiation with light in the F band. (**b**) Differences in VA caused by irradiation at $T = 80$ K in the F band (1) and after heating up to 160 K (2). Spectrum (3) was obtained for a sample with less OH^- doping

of these peaks are the drastically shifted transition energies and the strength of the lines. Since the trapping of electrons and the corresponding appearance of the new OH^--related defects reduce the number of regular OH^- defects, a reduction of the VA from the latter is expected. Indeed, a rather small negative signal is found if the difference between the spectra before and after light irradiation is examined (Fig. 8.3b, spectrum (1)). From the known VA positions we find that both OH^- and the $(OH^-)_2$ pair participate in the capturing of electrons. The correlation between these defects and the respective VA and EA absorption bands can be obtained by heating the sample above 160 K, a temperature at which one type of center disappears (Fig. 8.3, spectrum (2)). It turns out that at this temperature, only the electrons captured by $(OH^-)_2$ defect pairs are stable. Similarly, a variation of the OH^- concentration reduces the number of pairs drastically, so that only individual OH^- defects remain as possible molecular traps (see Fig. 8.3, spectrum (3)). Both methods yield a consistent assignment, indicated in Figs. 8.2 and 8.3. To reflect both the presence of a trapped extra electron and the trapping entity, we have labeled the center as $(OH^-)^-$ and $(OH^-)_2^-$.

8.2.2 Optically Detected Magnetic Resonance

Further insights into the nature of the defect were obtained by optically detected magnetic-resonance studies [12]. In the first step, the magnetically induced circular dichroism of the electronic absorption band was detected. Although the centers were unstable under illumination with the probe light, these measurements were made possible by simultaneous F light irradiation.

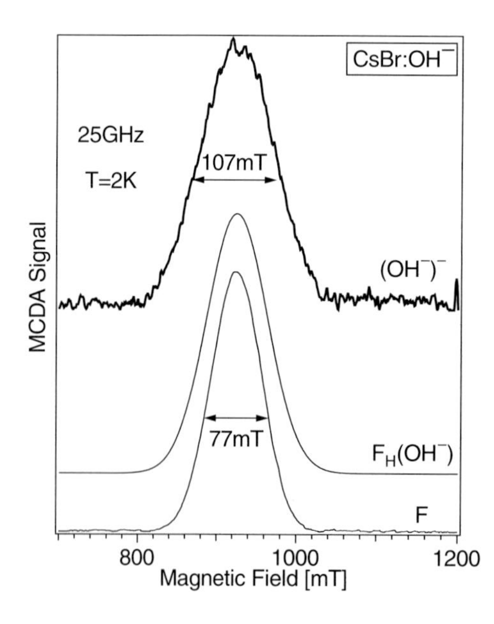

Fig. 8.4. Optically detected EPR spectra (MCD-EPR) of F center and of $F_H(OH^-)$ and $(OH^-)^-$ centers in CsBr

A tanh $(\beta B/kT)$ temperature dependence of the MCDA signal was found, which is typical of a spin-1/2 system. The one-electron picture underlying our trapped-electron model is further supported by the F-like g-factor of the excited state determined by T-dependent MCDA-measurements.

The MCDA signal can be used further to optically detect the EPR signal [13]. For direct comparison, the resulting spectra for the F center, the $F_H(OH^-)$ center, and the $(OH^-)^-$ centers are depicted in Fig. 8.4. The EPR band shows, in all cases an almost perfect Gaussian shape and has the largest width for $(OH^-)^-$, reflecting the delocalized character of the trapped electron and the resulting larger SHF interaction which occurs predominantly with the Cs neighbors. A similar result was obtained earlier by evaluating the absorption spectra.

8.3 Vibrational Properties of Molecular Electron Traps

In the identification of the molecular electron traps by VA in the previous section we have already noticed drastic changes (i.e. a shift and and change of strength) in the properties of the OH^- vibrations. These changes are of unique magnitude and hence it is worthwhile to consider them in more detail. An application of VA spectroscopy in which the differences of spectra before and after formation of the molecular electron trap centers are measured allows the detection of even very small absorption responses, such as those encountered for higher-harmonic transitions and librational sidebands, and thereby a sensitive determination of further important vibrational properties of this new type of molecular defect is possible. These properties are summarized in Table 8.1.

8.3.1 Shift in Transition Energy and Enhancement of Absorption Intensity

As can be seen in Table 8.1, all spectral positions of the $(OH^-)^-$ centers are shifted drastically to lower energy, far below the values for the free OH^- center, making the application of the simple model introduced in Sect. 3.1.2 with the (3.6) and (3.7) impossible. *The shifts cannot be explained just by changes in the repulsive interaction.* Owing to the closeness of the extra electron to the molecule the changes of interatomic O–H bond strength are more drastic. This assertion is further supported by the strongly enhanced transition strength which can be interpreted following the model of Sect. 3.2 in terms of the strong additional polarizability induced by the trapped electron. This electron is expected to be coupled sufficiently to the motion of the OH^- molecule and thereby adding considerably to the overall transition dipole moment. The dominance of the additional contribution to the overall absorption intensity is so strong that, in contrast to the isolated OH^- defect, the absorption is of almost identical strength for both isotopic variants measured,

Table 8.1. Parameters of the vibrational and librational modes of OH^-/OD^- ("1") and $(OH^-)^-/(OD^-)^-$("2") in Cs halides; $\overline{\omega}_i$ and I_i are the frequencies (in cm^{-1}) and absorption intensities of the transition i

Host		CsCl		CsBr		CsI	
Molecule		OH	OD	OH	OD	OH	OD
fundamental $\overline{\omega}_{01}$	1	3602	2655	3580	2642	3572	2636
	2	3149	2362	3262	2434	3358	2494
$\Delta\overline{\omega}_{10} = \overline{\omega}_{01}^{2-} - \overline{\omega}_{01}^{1-}$		453	292	318	208	214	142
I_{01}^{2-}/I_{01}^{1-}		20	> 100	8	60	4	20
$\Delta\overline{\omega}_{\mathrm{T}} = \overline{\omega}(100\ \mathrm{K}) - \overline{\omega}(5\ \mathrm{K})$	2	+7.2	5	7.5	5.3	5	3.5
second harmonic $\overline{\omega}_{02}$	1	7028	5216	6984	5188	6965	5178
	2	5903	4542	6204	4713	6448	4855
harmonic frequency $\overline{\omega}_{\mathrm{e}}$	1	3777	2749	3756	2738	3749	2731
	2	3544	2544	3583	2590	3626	2627
$\frac{\overline{\omega}_{\mathrm{e}}(\mathrm{OH})}{\overline{\omega}_{\mathrm{e}}(\mathrm{OD})}$	1	1.374		1.372		1.373	
	2	1.393		1.383		1.380	
$\Delta\overline{\omega}_{\mathrm{e}} = \overline{\omega}_{\mathrm{e}}^{2-} - \overline{\omega}_{\mathrm{e}}^{1-}$		233	205	173	149	123	103
anharmonicity shift $2x_{\mathrm{e}}\overline{\omega}_{\mathrm{e}}$	1	176	93	176	96	182	95
	2	395	183	321	154	264	133
anharm. parameter x_{e}	1	2.3%	1.7%	2.3%	1.7%	2.3%	1.7%
	2	5.6%	3.6%	4.4%	3%	3.7%	2.6%
$\frac{x_{\mathrm{e}}(\mathrm{OH})}{x_{\mathrm{e}}(\mathrm{OD})}$	1	1.377		1.34		1.39	
	2	1.54		1.51		1.43	
$\frac{I_{02}}{I_{01}}$	1	0.02%	0.86%	0.24%	1.9%	0.8%	5.4%
	2	4%	3%	7%	2%	16%	8%
librational frequency $\overline{\omega}_{\mathrm{libr}}$	1	279	207	288	212	348	255
	2	583	420	574	414	575	416
$\frac{\overline{\omega}_{\mathrm{libr}}(OH)}{\overline{\omega}_{\mathrm{libr}}(OD)}$	1	1.35		1.36		1.36	
	2	1.39		1.39		1.38	
$\frac{I_{\mathrm{libr}}}{I_{01}}$	1	25%	–	25%	30%	20%	18%
	2	6%	6%	15%	8%	16%	12%

OH^- and OD^-. Because the oscillator strength of isolated OD^- is lower in all cases studied, the enhancement factor is always higher for OD^- than for OH^-.

As shown in Fig. 8.5 where data for many spectral shifts and absorption intensities of F-electron-perturbed OD^- molecules are shown, the enhancement of the integrated absorption increases with increasing spectral shift. A similar

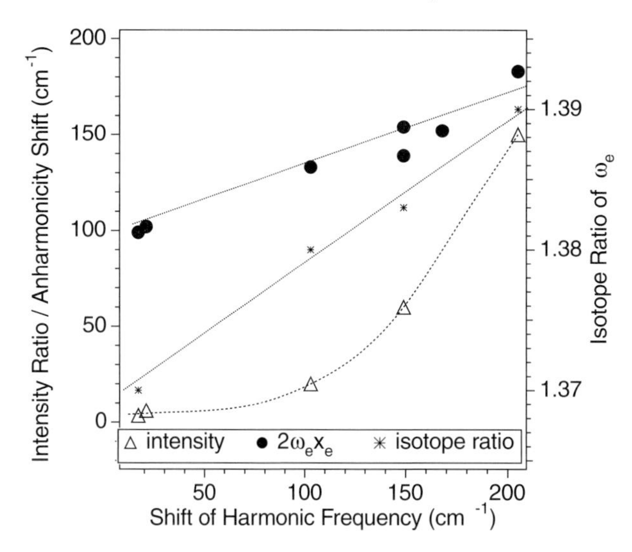

Fig. 8.5. Correlation of the frequency shift with the intensity enhancement I_{01}^{2-}/I_{01}^{1-}, the anharmonicity $2\overline{\omega}_{\mathrm{e}}x_{\mathrm{e}}$ and the OH/OD isotope ratio of $\overline{\omega}_{\mathrm{e}}$ in alkali halides. The data are plotted for $\mathrm{F_H(OH^-)}$ and $\mathrm{(OH^-)^-}$ centers in Cs halides (Tables 7.1 and 8.1), and for $\mathrm{(OH^-)_2^-}$ pairs in KCl and CsBr (Table 8.2). Compare similar plots for H-bonded compounds (e.g. Figs. 6 and 9 in Ref. [17])

correlation between the shift in the transition energy and the enhancement of the transition intensity has been found for hydrogen-bonded molecules in crystals and solutions [14, 15, 16, 17]. In that field, both quantities are used to quantify the strength of a hydrogen bond, and therefore intense experimental and theoretical investigations have been performed to account for these correlations [18]. Using various methods of quantum chemistry (SCF, CNDO, etc.; see e.g. [19]), it was found for water dimers for instance, that the weakening of the interatomic O–H bond is caused by the addition of electronic charge. Similarly, this shift of charge increases the transition probability through a *dynamic* modulation and polarization of the electric charge [20, 21]. This latter interpretation is essentially the same as the one that we have developed phenomenologically for our defect systems in the vicinity of a polarizable defect. Lacking quantum-chemical calculations for our electron trap centers, we can adapt the results for hydrogen-bonded systems and consider the combination of an $\mathrm{OH^-}$ molecule and a trapped electron as a quasi-hydrogen-bonded molecule. Within this model, we can picture the electron and the attractive potential caused by the lattice as an "atom", which makes a bond with the $\mathrm{OH^-}$molecule, i.e. some charge is transferred onto the molecule. This "bonding" becomes more pronounced as the lattice constant is reduced by variation of the host in the sequence CsI→CsBr→CsCl. Through a dynamic interaction between the $\mathrm{OH^-}$ vibration and the electron with participation of the host ions, the vibrational transition intensity can then be enhanced – as in water dimers – by a combination of polarization and charge redistribution effects. This analogy is further supported by the observation that the OH/OD isotopic ratio for the harmonic frequencies (see Fig. 8.5) is increasing with increasing energy shift, again just as found in hydrogen-bonded systems [22].

8.3.2 Mechanical and Electrical Anharmonicity

By evaluating the data for the higher-harmonic transitions using (3.3), it can be seen that the energy shift of the transitions is caused not only by a change of the harmonic eigenfrequency $\overline{\omega}_e$ but also by an up to twofold increase in the mechanical anharmonicity. As above, a clear correlation with the other changes exists. The drastic increase of x_e indicates a strong change in the shape of the anharmonic potential. Assuming a Morse potential for both the OH^- and the $(OH^-)^-$ molecules, the increased anharmonicity will result, according to (3.5), in a decrease in the dissociation energy U_D making the potential shallower. The deviations of the OH/OD isotopic ratios for the anharmonicity parameter x_e are similarly strong. This indicates not only that the OH/OD molecule is participating in the vibration but also that the electron distribution responsible for the intraatomic potential is not independent of the isotope.

In contrast to these drastic effects, the electronic anharmonicity, which is very pronounced and complicated for isolated OH^- defects, is much less important for the $(OH^-)^-$ center, such that the intensity ratios of the fundamental to the second-harmonic absorption are much closer to the value expected for the purely mechanical case, i.e. $I_2/I_1 = x_e$. It seems as if the mechanical picture of a simple "vibrating charged particle" is applicable much better to the electron trap center. Within the model developed above, the vibrating charge is not the regular charge of the OH^- molecule but the additional charge. The latter yields the dominating contribution to the transition probabilities.

8.3.3 Librational Sidebands

While for CsBr and CsI the librational sidebands in the VA spectra of the isolated OH^- molecule can be accounted for by a librational motion around the center of mass that is restricted by a high rotational barrier, the librational/rotational behavior is more complex for CsCl and is unique. In this host, various modes can be detected as sidebands, which belong to (1) a hindered rotational motion of a strongly dressed and/or largely off-center OH^- molecule,[2] (2) a libration around the center of mass, and (3) a radial oscillation of the off-center molecule [23]. For our comparison with the librational modes of the $(OH^-)^-$ center we chose only the librational motion around

[2] CsCl is the only ionic solid in which direct evidence for a rotational motion of the isolated OH^- molecule could be obtained experimentally.

the center of mass, because this was the only motion observed for the electron trap centers. These modes are characterized by a large isotope effect on their frequency. In a simple torsional harmonic oscillator model [24] a ratio $\frac{\overline{\omega}_{10}(OH)}{\overline{\omega}_{10}(OD)} = 1.375$ is expected and indeed, for the isolated molecules a value just slightly smaller is observed. For the $(OH^-)^-$ centers this ratio is close to a value of $\sqrt{2}$. The latter value is expected for a motion of the hydrogen with a fixed oxygen. As can be seen from the much higher librational frequency, the rotational barriers are made higher by the electron capture.

8.3.4 Summary

In summary, our experimental observations on the vibrational properties are consistent with the picture of a strongly orientationally aligned OH^- defect which by its off-center position creates in cooperation with the adjacent host ions, an attractive potential for an extra electron which is then able to share the anion vacancy. Owing to the small size of the molecule, there is enough "room" in the vacancy to ensure that electron repulsion is not strong enough at all locations to compensate the other attractive energies, making the arrangement described here an energetically stable configuration. It is apparent that the ability to create an attractive potential well for the electron will be very sensitive to the lattice constant and the host ion lattice. We find that when the host material is varied in the sequence CsCl→CsBr→CsI, the probability of forming an electron trap center increases, while the effects of the electron–molecular interaction decrease.

8.4 Electron Trapping by OH^- Pairs

We have concentrated so far on the properties of an individual OH^- defect perturbed by a trapped electron, but as we have already seen above, OH^- pairs are able to capture electrons as well, and even do so with higher thermal stability and better localization of the electron. Moreover, our observations have so far been limited to Cs halides.

The reason for the latter restriction becomes obvious if we consider the changes in the VA after photoionization in KCl:OH^- and KCl:OD^- shown in Fig. 8.6 in which the single extra VA line can be assigned on the basis of the observed reduction in the $(OH^-)_2$ absorption, to $(OH^-)_2^-$. Despite numerous attempts under various conditions, no extra VA line could be found which could be correlated with the trapping of electrons by isolated defects. Obviously, the attractive potential is too shallow in this case to allow a stable formation of the $(OH^-)^-$ center. The situation is similar in KBr. In the latter host material, two types of OH^- defect pairs have been found [25]: a (110) pair and a (200) pair. Only the first, however, can capture an electron, as far as we can observe.

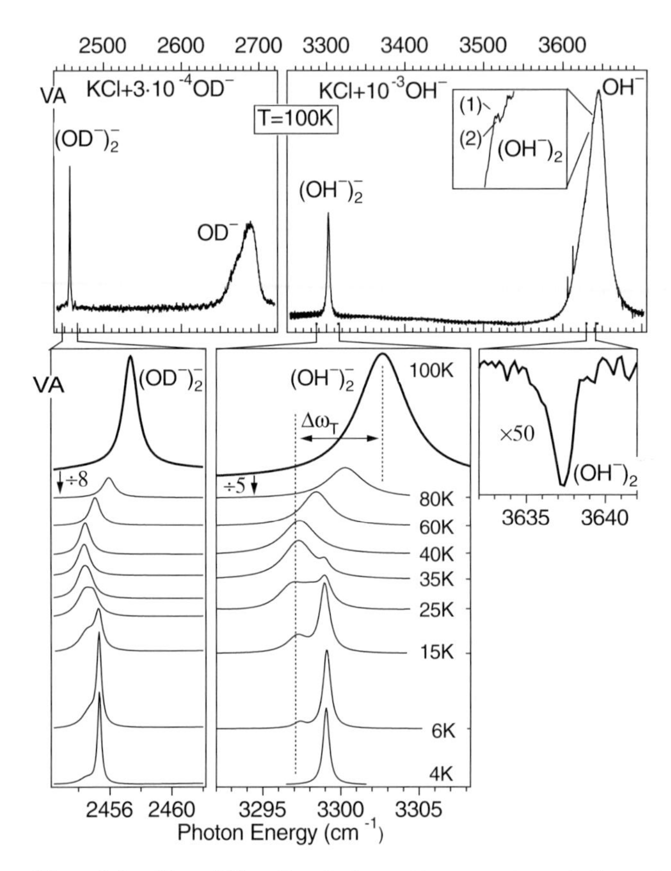

Fig. 8.6. *Top*: Vibrational-absorption spectra of F centers in KCl:OH^- and KCl:OD^-. *Bottom*: VA differences and their variation with temperature

In CsBr, in which both $(OH^-)_2^-$ and $(OH^-)^-$ centers can be studied it is found that the pair produces a more attractive and deeper potential for the electron. This is reflected in a higher temperature up to which the defect is stable, and by an electronic-absorption band which narrower and located at higher energies. All vibrational properties of the $(OH^-)_2^-$ pair electron trap centers are summarized in Table 8.2. In samples containing both isotopic OH^- and OD^-, not only pairs of the same species, i.e. $(OH^-–OH^-)^-$, but also mixed pairs $(OH^-–OD^-)^-$ with slightly different transition energies, can be observed.

Similarly to the situation for the $(OH^-)^-$ centers the vibrational properties of the $(OH^-)_2^-$ centers are drastically changed in all respects by the trapping of the electron. The more complicated structure of the center, however, makes it even harder to give a quantitative theoretical explanation of this behavior. The result of temperature variation show that the defect complex

Table 8.2. Vibrational parameters of $(OH^-)_2^-$ and $(OD^-)_2^-$ in KCl, KBr, and CsBr. Except for the ratios, all values are given in cm^{-1}

Host		KCl		KBr		CsBr	
isotopic species		$(OH^-)_2^-$	$(OD^-)_2^-$	$(OH^-)_2^-$	$(OD^-)_2^-$	$(OH^-)_2^-$	$(OD^-)_2^-$
$\overline{\omega}_{01}$	regular pair	3297.5	2454.6	3366.2	2499.8	3292.5	2448.4
	mixed pair	3305.8	2461.5	3372.2	2504.3	3294	2447.8
$\Delta\overline{\omega}_{10}$		343	225	251.4	169	288	193
$\Delta\omega_T$		6.2	3	2.4	1.2	5.9	4
$\overline{\omega}_{20}$		6277	4757	–	–	6299	4758.1
$\overline{\omega}_e$		3616	2607	–	–	3579	2587
$\Delta\overline{\omega}_e$		201	168	–	–	178	149
$2x_e\overline{\omega}_e$		318	152	–	–	286	139
x_e		5.5%	3%	–	–	4.5%	2.5%
$\frac{I_{02}}{I_{01}}$		3%	–	–	–	–	2%

in KCl besides consisting of more than one molecular defect, exists in at least two different configurations. As shown in the VA spectra in Fig. 8.6, the low-temperature configuration coexists at $T \approx 25$ K with a "high-temperature" configuration, which is the only one present for $T > 50$ K. This behavior is apparent for both isotopes and can be found in KBr as well.

8.5 Electron Tunneling from F Centers to OH^--Related Defects

The studies of electron trapping by OH^--related defects presented so far have been limited to low OH^- concentrations such that no F luminescence quenching occurred ($[OH^-] < 10^{-3}$ for KCl and $< 10^{-4}$ for Cs halides). In these samples the $(OH^-)^-$ and $(OH^-)_2^-$ centers could be produced only at relatively high temperatures, at which free electrons are created by thermally activated photoionization. The situation changes as the concentration is increased. In the same concentration range, in which the F center luminescence is quenched, it becomes possible to produce molecular electron trap centers and to bleach the F band even at the lowest temperatures.

This coincidence in the concentration-dependence is illustrated for CsBr in Fig. 8.7. In the samples with higher OH^- concentrations, the VA spectra also become more complicated. An example of this is shown in Fig. 8.8 for KCl. A large number of different OH^--related electron trap centers appear, some of which are not stable even at the lowest temperatures and can only be observed under simultaneous light irradiation. All these are possible candidates for the traps involved in the electron tunneling and the E–V transfer. However, the

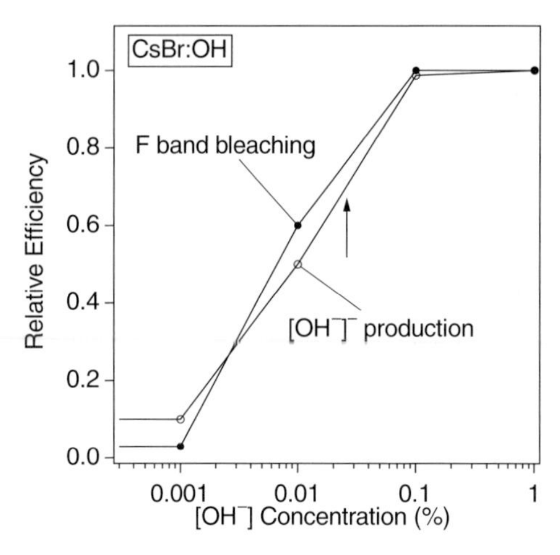

Fig. 8.7. Comparison of $(OH^-)^-$ center production yield and F band bleaching efficiency at $T = 4$ K as a function of OH$^-$ concentration. The production yield is given as the ratio of the number of $(OH^-)^-$ centers produced by light irradiation at $T = 4$ K to that at $T = 80$ K. The F band bleaching efficiency is calculated as the ratio of the optimal F band reductions produced at $T = 4$ K and 80 K. The critical concentration for F luminescence quenching is indicated by an *arrow*

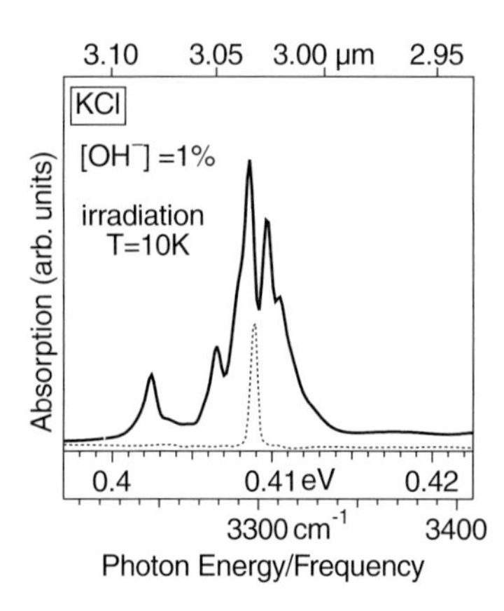

Fig. 8.8. OH$^-$-related electron trap centers produced by F band irradiation at low temperatures in highly OH$^-$-doped KCl. For comparison, the VA of an $(OH^-)_2^-$ center produced through thermally activated photoionization in low-doped KCl is shown by a *dashed line*

time resolution of the VA measurements was very limited, so that no electron back-tunneling faster than 30 s could be resolved, far away from the required rates of < 1 μs.

8.6 E–V Transfer Between Distant F Centers and OH$^-$Defects

Direct evidence that E–V transfer occurs and that vibrational states of the molecule get populated after F excitation can be obtained besides by VL also by the detection of anti-Stokes Raman scattering (ASR).[3] While both methods have been used successfully for the F–CN^- pair, only ASR is suitable for OH^- molecules owing to their strongly nonradiative decay. ASR studies of KBr have shown that E–V transfer occurs not only for the next-nearest-neighbor aggregate, the $F_H(OH^-)$ center, but also for slightly more distant OH^- molecules [26]. In the latter case, the transfer becomes more efficient when the distance is reduced by partial center aggregation. While this is already an initial indication that the E–V transfer can occur over larger distances we shall discuss the situation in CsBr in the following. For this host the results are clearer and the distances between the defect partners are much larger [27].

We used CsBr samples doped with OD^- at a concentration of 2×10^{-4}, at which the F luminescence is partially quenched. We chose the OD^- isotopic species because it yields stronger Raman signals compared with OH^- owing to the lower vibrational frequency and the longer vibrational lifetime [26]. Before the experiment the sample was carefully annealed and rapidly quenched to low temperature in order to secure a statistical center distribution. The result of the experiment is shown in Fig. 8.9. This figure shows the anharmonicity-shifted ($v \rightarrow v - 1$) lines of the OD^- molecules excited to a vibrational level v. The origin of the transitions can be identified by comparison with the VA spectra as essentially unperturbed ("distant") OD^- molecules. Owing to the strong influence of the F center on the molecular eigenfrequency in case of the $F_H(OD^-)$ center we would expect a shift to be observed for slightly longer distances also, so that we are quite certain that the OD^- molecules responsible for the observed ASR effect are several lattice constants away. This conclusion is further supported by the observation that even with careful stepwise aggregation the ASR signal is not increased, unlike the case for KBr. The vibrational excitation is quite high for the spectra shown, but it decreases at lower pump power so we have concluded that the E–V transfer proceeds predominantly into the vibrational levels $v = 1$ and 2, as for KBr. All these observations are in very good agreement with the proposed model. The search for ASR signals, which could be connected to vibrationally excited OD^- molecular electron traps was not successful. We were only able to find Stokes-shifted Raman signals for the stable $(OD^-)^-$ and $(OD^-)_2^-$ centers in their vibrational ground states. Obviously, the center involved in the electron tunneling can only be observed after the back-tunneling of the electron. The Raman signal for the electron trap centers is larger that for isolated OD^-, suggesting that, in addition to the VA absorption strength, the Raman cross-section is also increased for these center types. However, a quantitative evaluation is not possible with the data obtained so far, because the effect could be at least partially due to resonance enhancement.

[3] The work described here was performed in collaboration with E. Gustin and M. Leblans in Professor Schoemaker's group at the Unversity of Antwerp.

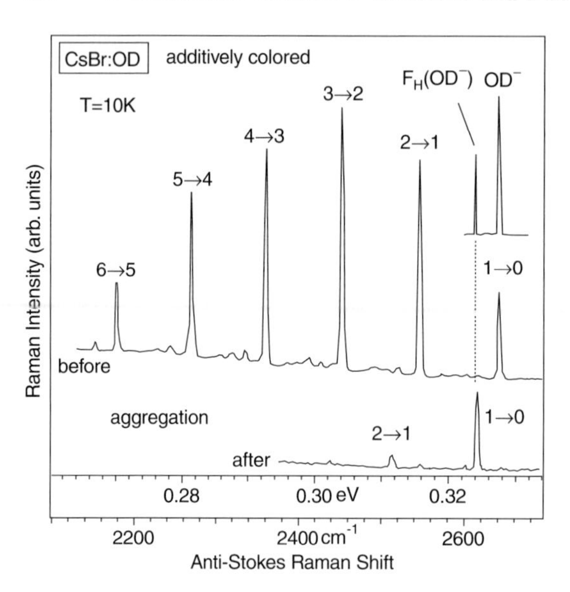

Fig. 8.9. Anti-Stokes Raman spectra of distant OH$^-$ defects vibrationally excited through F centers in a carefully quenched CsBr:OD$^-$ sample. For comparison, the VA spectrum of an aggregated sample is also shown

8.7 Conclusions and Outlook

Although we have been able to show a lot of experimental evidence that the electron transfer through tunneling and the E–V energy transfer are connected, a direct proof is absent. In other words, it cannot be excluded that the processes are not causally connected but instead occur in parallel. The incompleteness of the chain of evidence could be closed by time-dependent measurements in which the appearance of molecular electron trap centers is correlated in the time domain with the quenching of the luminescence and the appearance of ASRR lines. Measurements of that kind, however, are a very difficult task owing to the weak signals and the different spectral ranges. A more promising possibility is to study the changes in VA in pump–probe experiments using strong probe light sources, for instance optical parametric oscillators (OPO) which can be operated in pulsed and cw modes. Despite the final uncertainty, it is very attractive to adopt the model we proposed above as it explains the observed special properties of OH$^-$ in alkali halides quite naturally. Even the otherwise mysterious host lattice dependence of the critical doping level can be accounted for by simple statistics. While in Cs halides electron trapping and tunneling can occur for individual OH$^-$ molecules, in K halides only certain pairs or more complicated OH$^-$ related aggregates are suitable. These pairs are present only in small numbers, and therefore, even if the critical interaction distance ($R_{\mathrm{crit}} \sim 10$ lattice constants d) is similar to that in CsBr, higher OH$^-$ concentrations are required. On the basis of these considerations, the statistical model of Gomes and Luty [1], which takes account only of the interaction of the F center with isolated molecules and results in a smaller value for $R_{\mathrm{crit}} \sim 5d$, becomes questionable.

8.8 Further OH^--Type Centers in CsI

In CsI, further defect centers appear under F light illumination in samples in which $F \rightarrow F_H$ aggregation has been performed. Through various bleaching, center production, and quenching experiments, we were able to tentatively identify and spectroscopically characterize (see Table 8.3) the following center types:

- $F_H(OH^-)$ center;
- $(OH^-)^-$ center, discussed in Sect. 8.2;
- $(OH^-)^-$ next to an anion vacancy, i.e. F^{ion}–$(OH^-)^-$;
- F' center next to an OH^- molecular ion, i.e. F'–OH^-.

Schematic models of the centers are shown in Fig. 8.10. The vibrational frequencies of these centers are listed in Table 8.3. It is interesting to note that the latter two defect types can be formed after an electron is trapped by an $F_H(OH^-)$ center. The extra electron shares an anion vacancy either with the F center or with the OH^- molecule.

OH^- and OD^- molecules exhibit very short nonradiative lifetimes of the order of 10 ns [26] so that the radiative VL process has a very low quantum efficiency. Among all alkali halides studied the only case in which OH^- luminescence has been observed so far has been in CsI. For this host material Krantz and Luty [28] observed, after F_H center formation by means of excitation in the $F_H(1)$ band, a weak VL signal, which, however, did not coincide in its spectral position with the accurately measured VA band of either the $F_H(OH^-)$ center or the isolated OH^- molecule [23]. Comparison of the spectral positions observed in VA with the VL observed by Krantz and Luty [28] suggests that the F'–OH^- center may contribute to the VL.

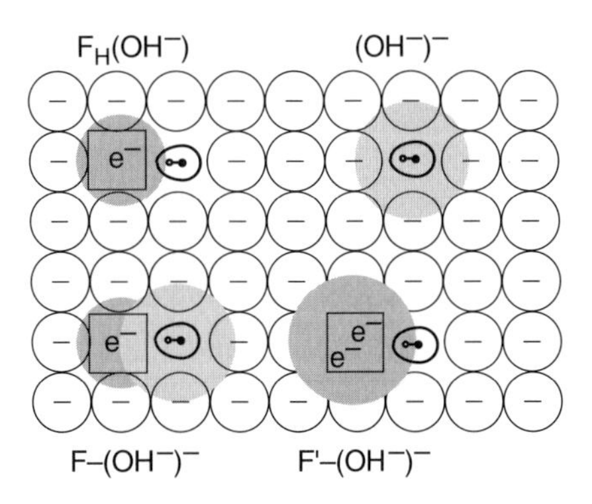

Fig. 8.10. Schematic models of various OH^--related defects in CsI after $F_H(OH^-)$ center aggregation

Table 8.3. Fundamental transition frequency and relative oscillator strength f_i/f_{isol} of various OH$^-$ defect centers perturbed by F electrons in various ways in CsI

		$F_H(OH^-)$	F'–OH^-	F^{ion}–$(OH^-)^-$	$(OH^-)^-$
$\overline{\omega}_{0\leftrightarrow 1}$ in cm^{-1}	OH	3542.8	3553.3	3338.8	3356.2
	OD	2616.4	2624.5	2478.5	2493.9
$\frac{f_i}{f_{isol}}$	OH	1	~ 10	–	4
	OD	3	50	8.5	20

This interpretation is supported by the strong enhancement of the transition probability for the F'–OH^- centers compared with all others (see Table 8.3). On the other hand, the shape of the VL spectra would be expected to depend strongly on the relative numbers of centers, which will be influenced by the "history" of the sample treatment and by the light irradiation. Nothing like this was observed and hence we have to admit that the nature of the VL in CsI is not fully understood at present.

References

1. L. Gomes and F. Luty, Phys. Rev B **30**, 7194 (1984).
2. H. Fedders, M. Hunger, and F. Luty, J. Phys. Chem. Solids **22**, 299 (1961).
3. F. Porret and F. Luty, Phys. Rev. Lett. **26**, 843 (1971).
4. I. Schneider, Solid State Commun. **25**, 1027 (1978).
5. E. Korovkin and T. Lebedkina, Sov. Phys. Solid State **33**, 1595 (1991).
6. M. Malghani and D. Smith, Phys. Rev. Lett. **69**, 184 (1992).
7. M. Malghani and D. Smith, Mater. Sci. Forum **239–241**, 365 (1997).
8. J. Vinti, Phys. Rev. **41**, 432 (1932).
9. V. Dierolf, H. Paus, and F. Luty, Phys. Rev. B **43**, 9879 (1991).
10. V. Dierolf, Master's thesis, Universität Stuttgart, 1987.
11. M. Krantz, Ph.D. thesis, University of Utah, 1987.
12. V. Dierolf and J.-M. Spaeth, Mater. Sci. Forum (Proc. ICDIM96) **239–241**, 461 (1997).
13. J.-M. Spaeth, J. Niklas, and R. Bartram, *Structural Analysis of Point Defects in Solids*, Springer Series in Solid-State Sciences, Vol. 43 (Springer, Berlin, Heidelberg, 1992).
14. D. Hadzi and S. Bratos, in *The Hydrogen Bond: Recent Developments in Theory and Experiments. Vol 2: Structure and Spectroscopy*, edited by P. Schuster, G. Zundel, and C. Sandorfy (North-Holland, Amsterdam, 1976), Chap. 12, p. 565.
15. D. Glew and N. Rath, Can. J. Chem. **49**, 837 (1971).
16. J. Lindgren and J. Tegenfeldt, J. Mol. Struct. **20**, 335 (1974).
17. L. Ojamäe and K. Hermansson, J. Chem. Phys. **96**, 9035 (1992).
18. J. Sauer, Chem. Rev. **69**, 199 (1989).

19. P. Schuster, in *The Hydrogen Bond: Recent Developments in Theory and Experiments. Vol 1: Theory*, edited by P. Schuster, G. Zundel, and C. Sandory (North-Holland, Amsterdam, 1976), Chap. 2, p. 27.
20. B. Zilles and W. Person, J. Chem. Phys. **79**, 65 (1983).
21. D. Swanton, G. Bacskay, and N. Hush, Chem. Phys. **82**, 303 (1983).
22. W. Mikenda, J. Mol. Struct. **147**, 1 (1986).
23. M. Krantz and F. Luty, Phys. Rev. B **37**, 7038 (1988).
24. M. Klein, B. Wedding, and M. Levine, Phys. Rev. **180**, 902 (1969).
25. A. Afanasiev and F. Luty, Solid State Commun. **98**, 531 (1996).
26. E. Gustin, M. Leblans, A. Bouwen, and D. Schoemaker, Phys. Rev. B **54**, 6963 (1996).
27. V. Dierolf, E. Gustin, D. Schoemaker, and F. Luty, J. Lumin. **76&77**, 526 (1998).
28. M. Krantz and F. Luty, Phys. Rev. B **37**, 8412 (1988).

9 Ytterbium Ions and CN^- Molecules

In this chapter, we turn our attention to the electronic states of rare-earth ions. These ions play an important role as luminescing ions in insulating materials and are often used as active ions in solid-state lasers. Their most common charge state is the trivalent positive one, which is unfavorable for incorporation into alkali halides. However, Eu^{2+}, Yb^{2+}, and Sm^{2+} also exist as divalent ions with high enough ionization energies to be stable. The most drastic interaction effects are found for the defect combination Yb^{2+} and CN^-, which we shall review in some detail in the following[1]. Compared with the F center systems, the coupling to the lattice is much weaker but still appreciable. In our context this represents the case of intermediate electron–phonon coupling.

Introducing the CN^- molecules as partner in a complex by codoping yields an interesting defect system, the only one in which all the interaction effects mentioned in Chap. 1 can be observed. Owing to favorable conditions (UV–VIS spectral range, long radiative lifetimes, etc.) an almost complete experimental characterization can be performed, constituting a solid base for theoretical studies; such studies have so far, however, been performed only in a fairly phenomenological way.

9.1 Crystal Growth and Sample Characterization

Yb^{2+}-doped crystals can be grown quite easily by either the Kyropolous technique or the Bridgman technique. Unless a divalent doping material is used, the trivalent Yb^{3+} supplied by the commonly used $YbCl_3$ or $YbBr_3$ doping material must be converted into the divalent form by a hydrogen gas treatment. For codoping with CN^- molecules, pure alkali cyanide powder (of the same cation as the host material) was used. In the Kryopolous method, this material can be added to the melt quite easily in one or two steps during crystal growth yielding for example an Yb^{2+}-doped crystal which contains in part A no CN^- and in parts B and C increasing amounts of codoped CN^-. In

[1] Many of the experimental results and interpretations presented in this chapter are the product of a collaboration with Professor F. Luty and Dr. C.P. An at the University of Utah [1].

such samples, the defect interaction becomes apparent to the naked eye: the appearance of the crystal changes from completely colorless to an increasingly yellow coloration as one passes from part A to C, and this is matched by the colors that can be seen already in the melt. Similarly, in Bridgman-grown samples, the Yb^{2+} and CN^- doping gradient along the growth direction is reflected by a change in color from yellowish to orange (Fig. 9.1).

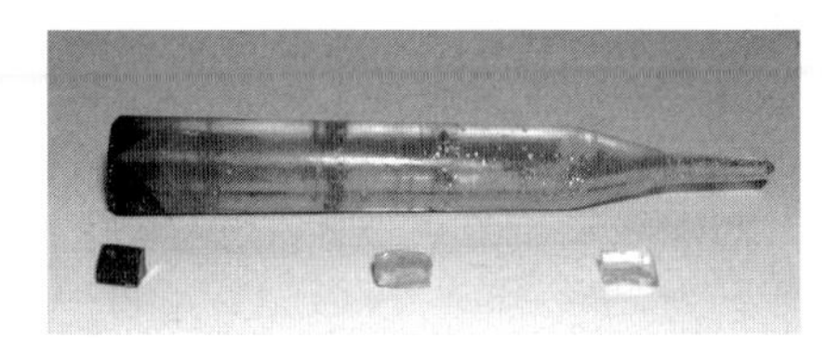

Fig. 9.1. Bridgman-grown KCl crystal doped with 3% KCN and 0.2% $YbCl_3$. Small pieces cleaved out of an identical crystal are shown below the crystal (dark grey on the left of the crystal corresponds to a dark orange and the lighter grey in the center to a bright yellow)

In the extreme case of pure KCN host doped with Yb^{2+}, both the melt and the resulting crystals are very opaque and have a strong brown color. These observations reflect the fact that Yb^{2+} and CN^- ions interact strongly, independently of whether they are in a crystal or in the melt. Further evidence that they are grouped together in the melt comes from the observed differences in the amounts of dopants which end up in the crystal. The "distribution coefficient" (i.e. the ratio of the concentrations in the crystal and melt) obtained from doping with Yb^{2+} alone varies strongly with the size of the cation of the host, having values of 0.2 for NaCl, 0.05 for KCl, and 0.01 for RbCl. In the latter two hosts the Utah group found considerable increases of those values (by up to a factor of 4) when the melt was codoped with increasing amounts of CN^-. In NaCl, the concentration of isolated Yb^{2+} ions relative to that of Yb^{2+}:$(CN^-)_n$ complexes is considerably higher than in the other host materials. The reasons for this defect-type-dependent distribution coefficient are the following:

- As Yb^{2+} is nearly equal in ionic size to Na^+, isolated Yb^{2+} ions can enter the solid from the melt quite easily for NaCl, but are rejected much more strongly in K^+ and Rb^+ halides owing to "size misfit".
- Yb^{2+}–CN^- pairs which already exist in the melt enter into the solid best when the size of the pairs matches to that of two neighboring lattice ions. This leads for example to a higher distribution coefficient for pairs approximately four times higher than for isolated Yb^{2+} ions in a RbCl host.

9.2 Yb^{2+}:$(CN^-)_n$ Defect Complexes: Electronic Transitions

In contrast to the very small variation of the absorption and emission spectra of Yb^{2+} in four different hosts (Chap. 2), drastic changes occur on substitution of host anions by CN^- molecular ions.

9.2.1 Absorption and Emission Properties

In Fig. 9.2 the absorption and emission spectra of Yb^{2+} in $(KCl)_{1-x}$:$(KCN)_x$ hosts with variation of x in four steps from $x = 0$ (pure KCl) to $x = 1$ (pure KCN) are presented. For easy comparison, each measured spectrum is normalized to an Yb^{2+} concentration of $\sim 10^{-4}$. For $x = 10^{-3}$, new, weak and quite broad Yb^{2+}-related absorption bands appear at the low-energy side of the regular A absorption band. Optical excitation of this sample at $\sim$ 310 nm produces the "blue" A* emission at $\sim$ 450 nm typical of isolated defects (shown in Fig. 2.3) indicating that they are still the dominant center type. On the other hand, excitation at $\sim$430 nm within the new absorption bands yields a new emission ("orange") with a maximum at $\sim$ 580 nm, which is related to a small number of Yb^{2+} ions perturbed by CN^- ions. Assuming that the complexes involve only a single CN^- ion and that the total oscillator strengths of the absorption remain unchanged, one finds an estimated concentration ratio of $[Yb^{2+}$–$CN^-]/[\text{isolated } Yb^{2+}]$ slightly higher than expected from statistics.

The new emission bands are not mirror images of the corresponding absorption. This observation indicates that, after optical excitation in the al-

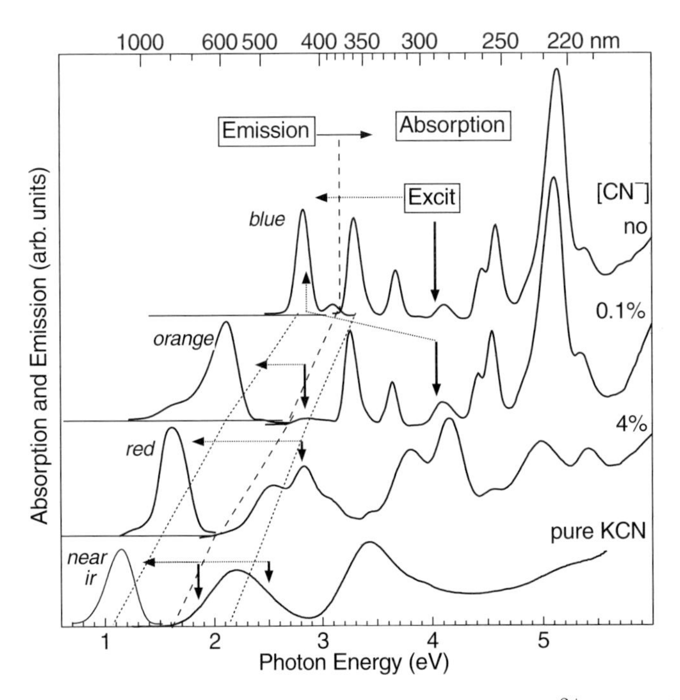

Fig. 9.2. Absorption and emission spectra of Yb^{2+}-doped $(KCl)_{1-x}$:$(KCN)_x$ for $[CN^-]$ concentrations x ranging from 0 to 100%. The excitation wavelengths are indicated by *arrows*. All data were taken at 15 K except for the data for pure KCN which were measured at 190 K owing to the strong light scattering by elastically ordered domains below 168 K [2]

lowed absorption band (A-type absorption), a relaxation takes place into a level from which transitions are forbidden or only weakly allowed similarly to the situation for isolated defects. This results in an A*-type emission with rather long radiative lifetimes (of the order of milliseconds) at low temperatures. Unlike the situation for the isolated defects, this relaxation behavior is almost identical for all host materials studied. Moreover, the radiative lifetime (as shown in Fig. 9.30) increases much more drastically with increasing temperature. This interesting observation may become the key to several applications and will be discussed later on.

As x increases to 0.04, the new absorption bands become stronger and extend to even lower energies. Optical excitation of the sample with $x = 0.04$ on the low-energy side produces a "red" emission at $\sim$ 780 nm. In this CN^- concentration range the system is strongly disordered, and inhomogeneously broadened absorption and emission bands are expected. While this behavoir is not easily seen in the regular absorption (Fig. 9.2), excitation (Fig. 9.3b) and emission spectra (Fig. 9.3c), because the contributions of the individual center

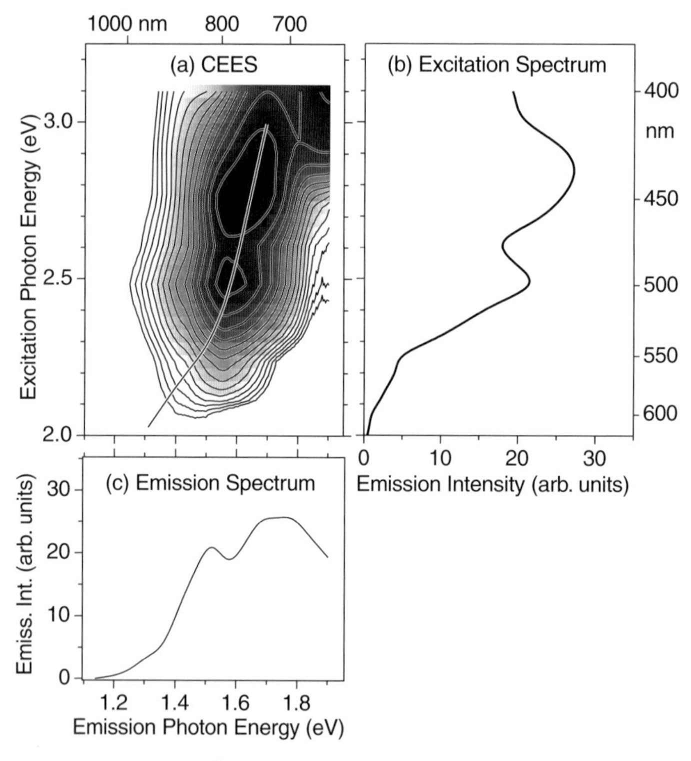

Fig. 9.3. $KCl:Yb^{2+}$+ 4% KCN: (**a**) Contour plot of CEES measurement at $T = 10$ K; (**b**) excitation spectrum of the total emission obtained from (a) by summing in the horizontal direction; (**c**) total emission spectrum obtained by summing in the vertical direction

type are obscured, the inhomogeneous character becomes very obvious in the combined excitation–emission spectrum. In the contour plot (Fig. 9.3a) of the CEES measurement for KCl:Yb^{2+}:CN^-, it is clearly observable that with a decrease of excitation energy the emission response is shifted continuously towards lower energies. Drawing a line perpendicular to the constant-intensity lines of the contour plots (i.e. "along the ridge of the mountain range") gives a curve which is almost vertical for high excitation energies and becomes diagonal (as for Raman responses) for lower energies. This can be interpreted as follows. Each of the various center types exhibits two absorption peaks in the spectral region depicted but the emission is governed at the high-energy side by the dominant (most abundant) types. Only when the dominat types not absorb anymore do other centers with lower transition energies become apparent. These latter centers show an almost equal shift of the emission and the lowest-energy absorption.

Turning our attention from this disordered case of KCl:KCN mixed crystals to the pure KCN host, in which every Yb^{2+} defect is surrounded by $n = 6$ CN^- neighbors, we find that the most drastic spectral changes are observed here: The lowest-energy absorption band has a peak position at 1.6 eV and the emission is centered at around 1.1 eV. Overall, we find a clear trend in which the redshift of the absorption and emission increases with the concentration of CN^- ions and hence with the number n of CN^- neighbors of a Yb^{2+} ion.

Despite these drastic changes induced by CN^- doping, the spectral behavior of the Yb^{2+} ion with increase of the CN^- concentration is almost identical in all host materials studied, indicating that the Yb^{2+}–(CN^-) interaction dominates over the interaction of the Yb^{2+} ion with the lattice.

For a more systematic analysis of the spectral shifts and their correlation with the various defect types, the measured absorption and emission spectra were decomposed into individual spectra of Yb^{2+}:$(CN^-)_n$ complexes with particular value of n using the excitation spectra of the characteristic emissions and the changes of absorption caused by thermal treatment. As an example, the results for the KCl host material are shown and discussed in the following.

Figure 9.4 shows the excitation spectra probed in the characteristic spectral regions "blue" ($\sim$ 450 nm), "orange" ($\sim$ 550 nm), and "red" ($\sim$ 750 nm) indicated in Fig. 9.3. While the spectrum for the blue emission clearly follows the absorption spectrum of the isolated Yb^{2+} centers (best visible for the **A** and **B** absorption bands), the spectra for the orange and red regions are drastically different from the latter as well as from each other. Therefore it is clear that they represent different Yb^{2+}:$(CN^-)_n$ defect types. We have made, intuitively, the following assignment: "blue" $n = 0$, "orange" $n = 1$, "red" $n \geq 2$ which will be discussed and justified below.

Thermal treatments were used to partially convert one defect complex into another, providing further useful information for the decomposition of

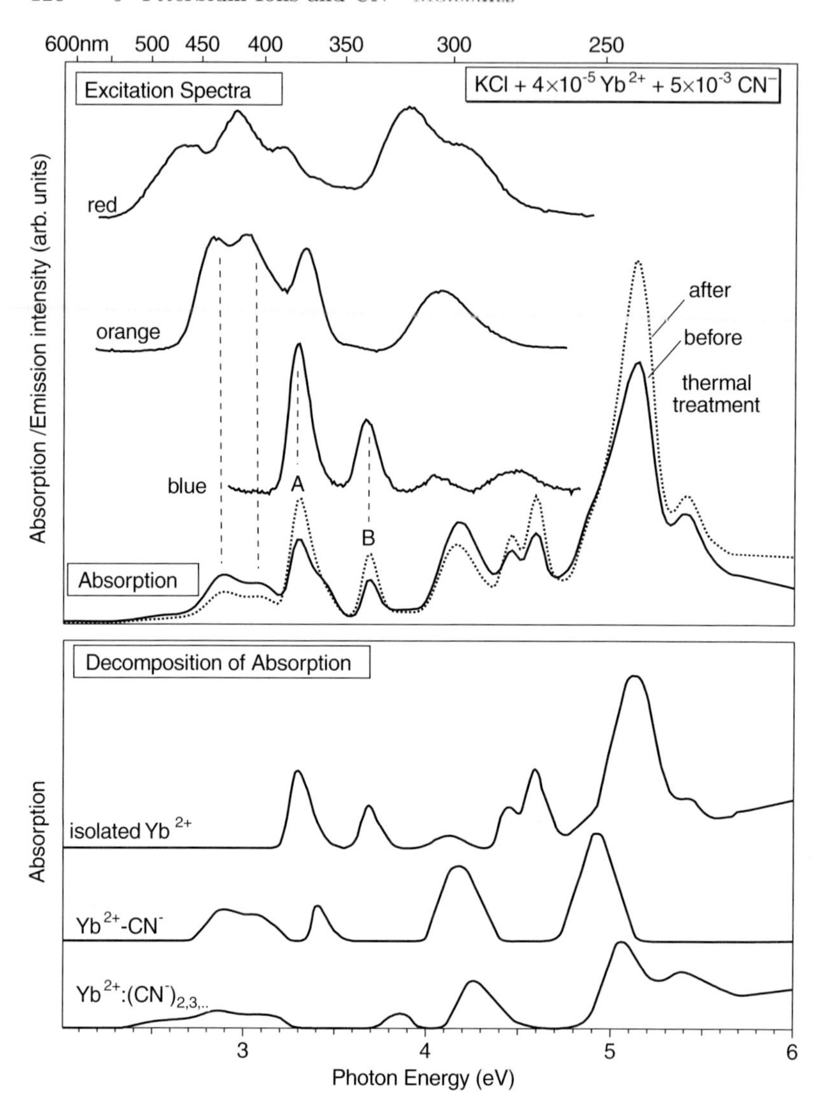

Fig. 9.4. *Top*: Excitation spectra probed in the blue, orange, and red emission regions. For comparison, the absorption spectra before and after a quenching procedure are shown. *Bottom*: Decomposition of the absorption spectrum on the basis of the measurements above into contributions from isolated Yb^{2+}, simple Yb^{2+}–CN^- defects and more complex Yb^{2+}:CN^-defects

absorption spectra. One type of sample ("before") consisted of crystals which were slowly annealed from the crystal melting point to room temperature just after crystal growth and subsequently stored for some time (a few weeks) at room temperature. The measured low-temperature absorption spectra of these samples (Fig. 9.4, top) were compared with spectra obtained just after fast quenching from $\sim 150\,°C$. For the KCl sample with a concentration of Yb^{2+} of 4×10^{-5} and a concetration of CN^- of 5×10^{-3}, the absorption of isolated Yb^{2+} increases, while the absorption of more complex defects decreases. However, the total integrated absorption intensity does not change after thermal quenching.

Using the excitation spectra and the results of thermal treatments, the absorption spectra can be decomposed into the contributions to the absorption from three groups, with $n = 0$, $n = 1$, and $n \geq 2$ of Yb^{2+}:$(CN^-)_n$ defect complexes. The spectra obtained (Fig. 9.4, bottom) are in good agreement with the blue, orange, and red excitation spectra, respectively (Fig. 9.4, top). For the simple Yb^{2+}–CN defect pair, the EA spectra consist of several bands and strongly resemble the spectrum for isolated Yb^{2+} defects (bands **A**–**F**), just shifted to lower energies by ~ 0.35–0.5 eV. Further shifts of the complete spectrum are found for the higher n values ($n \geq 2$), although a distinction between the absorption contributions of the complexes with $n = 2$ and ($n \geq 2$) is very difficult. A crude estimate can be obtained by comparing the decomposed absorption spectra for samples with high and low CN^- concentrations. The transition energy of Yb^{2+}:$(CN^-)_2$ shifts by ~ 0.35 eV compared with that of the Yb^{2+}–CN^- pair, while the absorption bands of the more complicated complexes ($n \geq 3$) shift even further to lower energies. As the endpoint of the variation of n, we take the defect complex Yb^{2+}:$(CN^-)_n$ measured directly without decomposition for pure KCN. The lowest-energy band of its electronic absorption is located at ~ 1.6 eV (see Fig. 9.3), which is about 1.5 eV lower than the **A** absorption band of isolated Yb^{2+} ions ($n = 0$). Although the average shift of A-type transitions is about 0.25 eV (≈ 2000 cm^{-1}) per neighbor, it decreases for increasing n such that the absorption bands belonging to Yb^{2+}:$(CN^-)_n$ defect complexes with $n = 3$, 4, 5, and 6 strongly overlap spectrally and cannot be decomposed. A similar conclusion about the spectral shifts can be drawn from the emission spectra, which exhibit, as shown in Fig. 9.3, a smooth (not stepwise) shift towards lower energies with decreasing excitation energy.

In summary, it has been shown in this section that:

- The electronic transitions of the Yb^{2+} ion in alkali halides are shifted to lower energies owing to the perturbation from CN^- neighbors.
- The shift increases with the number n of neighbors.

From the thermal-treatment experiments, we obtain further the following important information:

- The defect complexes Yb^{2+}:$(CN^-)_n$ can be converted partially into simpler forms by a thermal treatment at ~ 150 °C (a thermal energy of about 35 meV). Obviously, the binding energy between the Yb^{2+} and CN^- ions in a KCl host is rather weak, and is similar to that of pairs of rare-earth defects in alkali halides [3].

- The integrated oscillator strengths of Yb^{2+} ions are almost equal for the isolated form and for the defect complexes Yb^{2+}:$(CN^-)_n$, as was assumed above.
- A slow aggregation of centers takes place, forming preferentially $(CN^-)_n$-related Yb^{2+} complexes already at room temperature.

These properties are very similar to those of F center systems, raising the hope controlled photoinduced center aggregation of Yb^{2+}:$(CN^-)_n$ centers analagous to that of F centers should be possible. Unfortunately, all attempts in this direction have been unsuccessful so far.

9.3 Vibrational Transitions of CN^- Molecules Within Yb^{2+}:$(CN^-)_n$ Complexes

After concentrating, so far, on the changes in the Yb^{2+} electronic transitions due to the presence of CN^-, we turn our focus now to the changes in the vibrational transitions of a CN^- molecule due to perturbation by a Yb^{2+} ion located in its vicinity. In Fig. 9.5 several vibrational absorption spectra of CN^--related defects in KCl are shown. Starting in Fig. 9.5a with the low-CN^--doped sample, we find, besides the well-known absorption band of the isolated CN^--ion with the librational sidebands that are typical in KCl, several sharp peaks which occur only in Yb^{2+} samples (except for the peaks originating from CN^-–CN^- and Na–CN^- pairs). The peaks in question can be identified by concentration variation. In the case depicted (Fig. 9.5b) the concentrations of Yb^{2+} and CN^- were chosen such that the product $[CN^-]\times[Yb^{2+}]$ remained approximately constant, and with it the statistical probability of forming the simple Yb^{2+}–CN^- pairs. Consequently, the five lines marked L, which hardly change under this variation, can be assigned to the simple defect pair. Similarly, we find several lines (H) which occur only for high Yb^{2+} concentrations and which are most likely to be due to CN^- molecules incorporated into small $(Yb^{2+})_n$ aggregates. Their relative concentration can be reduced by carefully quenching the sample before the measurements.

If we increase the CN^- concentration above a few percent (Fig. 9.5c), all absorption lines broaden because of the inhomogeneity induced in the sample by the disordered elastic CN^- dipoles [4]. The measurement is complicated by strong "background" from the isolated CN^- ions. Nevertheless, by careful measurements and comparison with a sample doped with CN^- only several Yb^{2+}-related bands could be found and we assigned them to Yb^{2+}:$(CN^-)_n$ complex centers. In comparison with the isolated CN^- defect, the spectral positions of the Yb^{2+}-related CN^- transitions, with only one exception, are all shifted to higher energies.

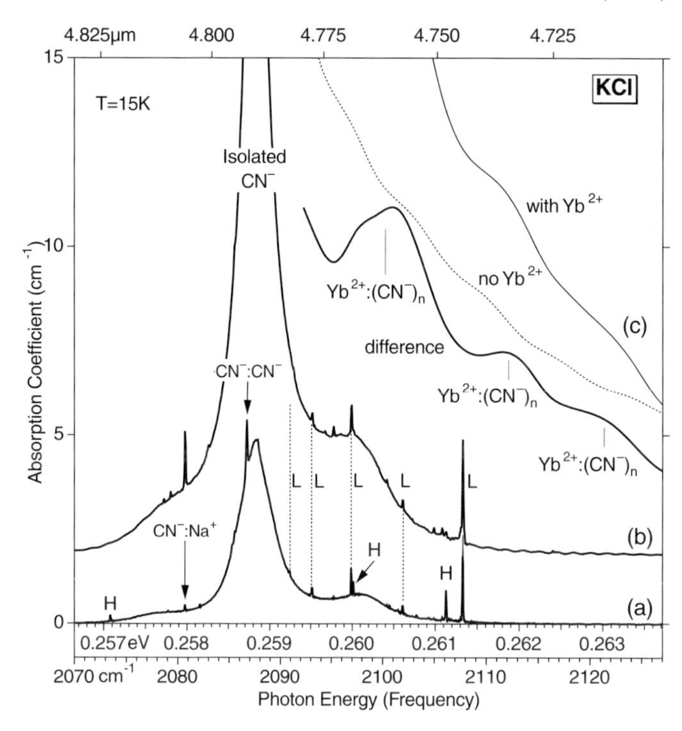

Fig. 9.5. Several vibrational absorption spectra for KCl doped with (*a*) $2{\times}10^{-4}Yb^{2+}+\ 10^{-3}CN^-$, (*b*) $4{\times}10^{-5}Yb^{2+}+\ 5{\times}10^{-3}CN^-$, and (*c*) with $8{\times}10^{-2}CN^-$ both with (10^{-3}) and without Yb^{2+}. In (*c*) the difference between the two spectra is also depicted. Several peaks are assigned to simple Yb^{2+}–CN^-defects (L), to $(Yb^{2+})_n$–CN^-(H), and to Yb^{2+}:$(CN^-)_n$. See text for details

9.3.1 Temperature Dependence

Turning our attention back to lower CN^- concentration, Fig. 9.6 shows the temperature variations of the VA spectrum. As is typical for orientationally aligned CN^- defects in alkali halides the initially sharp Lorentzian-shaped VA lines broaden and shift to lower energies [5]. In the simplest case this can be explained by a weak coupling of the vibration to a single (dominant) phonon mode [6]. The T dependence of the spectral position and line width can then be expressed by quite simple analytic expressions involving only the phonon frequency $\overline{\omega}_0$, the spectral width γ_0 (cm^{-1}) of the phonon band, and the coupling constant $\delta\overline{\omega}$ as parameters [7]:

$$\mathrm{FWHM}(T) = -\frac{2(\delta\overline{\omega}_0)^2 \mathrm{e}^{\frac{hc\overline{\omega}_0}{kT}}}{\gamma\left(\mathrm{e}^{\frac{hc\overline{\omega}_0}{kT}}-1\right)^2}\,, \tag{9.1}$$

$$\mathrm{Pos}(T) = \mathrm{Pos}(T=0\ \mathrm{K}) + \frac{\delta\overline{\omega}_0}{\mathrm{e}^{\frac{hc\overline{\omega}_0}{kT}}-1}\,. \tag{9.2}$$

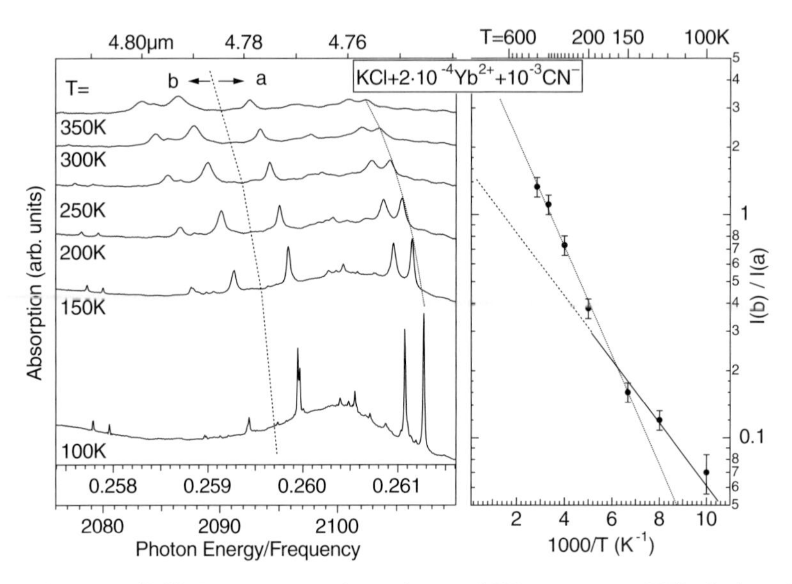

Fig. 9.6. (*left*) Temperature dependence of VA spectra in KCl. A *dotted* line separates the spectral regions of centers of type a and b. On the *right* ratio of the integrated intensities of type a and b centers is depicted. See text for details

In Fig. 9.7, the T dependence of the spectral width and position is shown for the most dominant band (indicated by a dotted line in Fig. 9.6), which remains well resolved over the whole temperature range. The solid lines represents fits to (9.1) and (9.2) with $\overline{\omega}_0 = 150\ \mathrm{cm}^{-1}$, $\gamma = 44.6\ \mathrm{cm}^{-1}$, and $\delta\overline{\omega}_0 = -4.5\ \mathrm{cm}^{-1}$. This parameter set achieves a fairly good agreement with the experimental data and the condition of weak coupling ($\delta\overline{\omega}_0 << \gamma$) is reasonably well fulfilled. Moreover, the values for the phonon frequency and width coincide closely with a peak in the phonon spectrum of KCl [8]. The discrepancy between the measured and predicted widths at low temperature

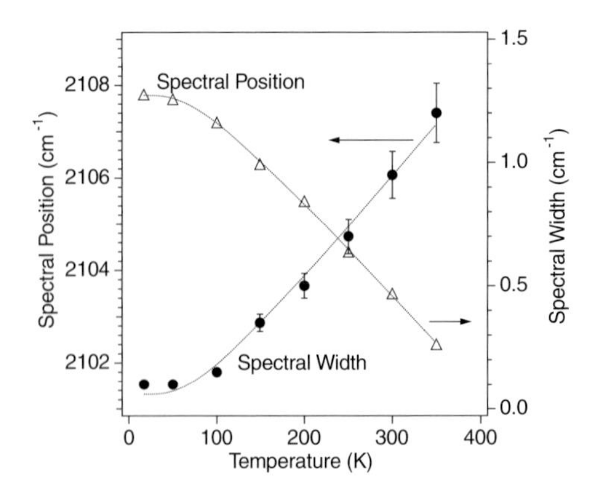

Fig. 9.7. Temperature dependence of the spectral position and spectral width of the vibrational transition of a Yb^{2+}–CN^- defect pair (indicated by a solid line in Fig. 9.6) in KCl

is due to the limited spectral resolution. The small coupling of the vibrational motion to the lattice observed here justifies the neglect of this coupling in all treatments of the energy transfer and interaction effects.

Besides the shift and broadening, a change and redistribution of the strength of the various absorption lines is observed. With increasing temperature the lines with higher energy (type a) are reduced, while the lower-energy lines (type b) become stronger. The spectral separation between the two groups is indicated in Fig. 9.6 by a *dotted* line. The trend can be seen more quantitatively on the right of the figure, where the integrated ratio of the strengths of the VA from centers of type b and a, $I(\mathrm{b})/I(\mathrm{a})$, is plotted versus temperature.

Fitting the data between $T = 150$ K and 350 K with a simple Arrhenius-type dependence yields an activation energy of ≈ 50 meV. However, a deviation is found at low temperatures. On the other hand, if the low-temperature data is fitted a lower activation energy is obtained (*dashed* line in Fig. 9.6, right) and the fitted value approaches $I(\mathrm{b})/I(\mathrm{a}) \approx 2$ as $\mathrm{T} \to \infty$. Obviously, we are dealing with a temperature-dependence of either the depth of the potential well which separates the two center configurations or of the integrated absorption strength ratio. The underlying bistable reorientation between the two center types which can also be induced optically is treated in Sect. 9.4.

Similarly to KCl, the VA spectra of NaCl samples codoped with Yb^{2+} and CN^- defects, shown in Fig. 9.8 (bottom), exhibit drastic changes and additional lines compared with the samples doped with only CN^-. However, the spectra are more complicated in several respects as follows.

Firstly, a broad VA band appears on the high-energy side of the band corresponding to isolated CN^-. This band depends strongly on the preparation of the sample. It is most pronounced in crystals which have been stored in the dark for several days. This suggests that in this host, the Yb^{2+} ions have a strong tendency to migrate through the sample at room temperature and preferentially form larger complexes with other Yb^{2+} ions and CN^- molecules. This behavior is well known for Eu^{2+} centers, which under certain conditions can form Eu-related subphases in NaCl [3]. The broad VA response almost completely disappears if the sample is heated to around 200 °C for 10 min and is subsequently cooled down to low temperature. The difference caused by this "quenching" procedure is shown by the shaded area underneath the absorption curve in Fig. 9.8. Despite the strong reduction of the broad absorption background, no similar increase of VA at other spectral positions can be found. Since we do not expect a decrease in the number of molecules, we have to assume that the transition probability for the defects associated with the broad VA response is stronger than that for the isolated and for the sharp Yb^{2+}-perturbed CN^- lines.

Secondly, there are many more rather sharp lines, which makes an unambiguous assignment of them very hard. We assume that in NaCl it is very favorable for the Yb^{2+} ions to aggregate forming complexes which involve

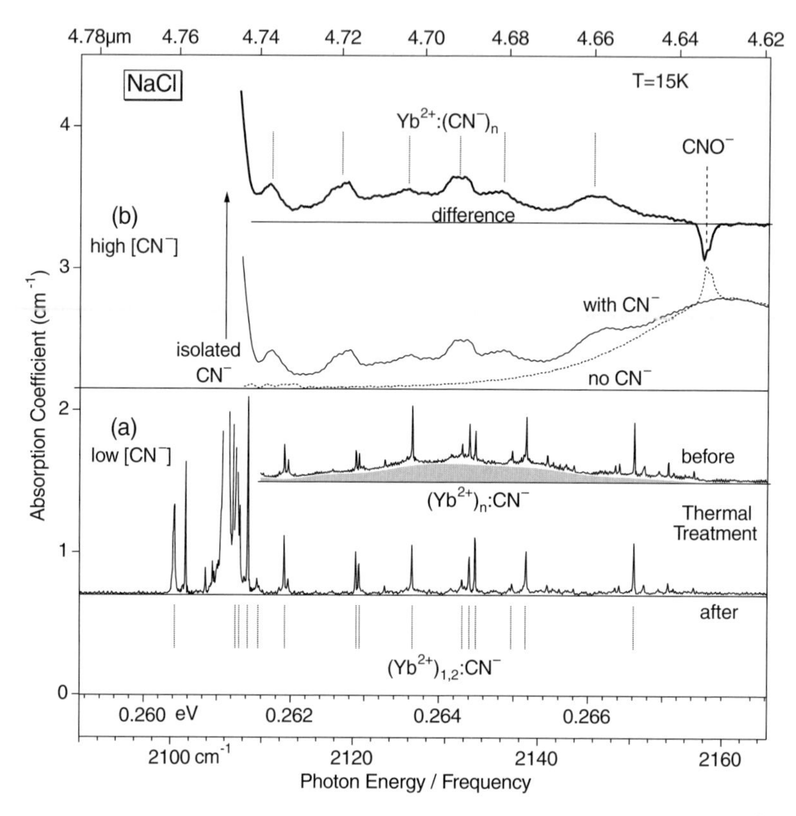

Fig. 9.8. Several vibrational absorption spectra for NaCl. (**a**) $3 \times 10^{-3} Yb^{2+} + 5 \times 10^{-4} CN^-$. Spectra before and after thermal treatment are shown. The gray area indicates the induced change. (**b**) $3 \times 10^{-2} CN^-$, both with and without Yb^{2+}(1.5×10^{-4}). The difference between the two spectra is shown as well. See text for details

a CN^- ion and more than one Yb^{2+} ion. By performing systematic variation of the sample preparation, light irradiation at low T, and temperature variation, we were able again to categorize the VA lines into two groups a and b, which are most likely related to each other in pairs ($a_i \leftrightarrow b_i$). As in KCl, a conversion between types a and b can be achieved by thermal activation. For higher CN^- concentrations, several additional broad VA transitions are found (see Fig. 9.8), which are related to Yb^{2+}:$(CN^-)_n$-complexes.[2] The results for RbCl and KBr are similar, so that the results of study of the vibrational absorption described above can be summarized as follows:

- The vibrational transitions of the CN^- ion in alkali halides are shifted to higher energies owing to the perturbation by Yb^{2+} neighbors.

[2] Besides showing clearly the Yb^{2+}-related VA responses of CN^-, the difference between a sample with and without Yb^{2+} illustrates that, as a side effect of Yb^{2+} codoping the often unavoidable CNO^- impurities disappear almost completely.

- For simple Yb^{2+}–CN^- pairs, the molecules are quite well orientationally aligned.
- The molecules are only weakly coupled to the lattice and other low-frequency modes.
- Very complex defect aggregates exist, which have a broad spectral response and show a drastically enhanced absorption transition strength; this is most apparent in NaCl.

The spectral data of all VA lines observed are summarized in Table 9.1.

Table 9.1. Fundamental transition energies $\overline{\omega}$ observed for various Yb^{2+} related CN^- centers in several alkali halides (given in cm^{-1}). The lines are ordered into different categories whenever the experimental findings have made this possible

NaCl

$(Yb^{2+})_{1,2}$–CN^-						$(Yb^{2+})_m$–CN^-	Yb^{2+}–$(CN^-)_n$	
type a			type b					
2150.4	2138.7	2126.4	2108.5	2107.6	2100.5	2133	2146	2136
	2133.3	2120.6	2112.5	2107.2			2131	2126
	2132.6	2120.3	2112.9				2119	2111

KCl

Yb^{2+}–CN^- pairs						$(Yb^{2+})_m$–CN^-	Yb^{2+}–$(CN^-)_n$
type a			type b				
a_1	a_2	a_3	b_1	b_2	b_3		
2107.8	2102.0	2097.0	2093.2	2090.8	2089.5	2106.2	2121
						2097.2	2112
						2073.5	2100

RbCl

$(Yb^{2+})_{1,2}$–CN^-					
type a			type b		
2099.1	2097.6	2092	2089	2087.7	2086.4
	2097.1	2091.2			2085.9
	2097	2092.6			

KBr

$(Yb^{2+})_{1,2}$–CN^-					
type a			type b		
2098	2096.4	2092.3	2086.4	2083.8	2082.8
		2091.3	2086.7		
		2090.8			

9.4 Optically Induced Bistability

The temperature dependence of the VA spectra has shown that different center configurations exist within the host materials studied which are energetically separated by a potential well. We have already encountered a similar situation for the $F_H(OH)$ centers in several host materials [9, 10, 11]. For that defect system, the barrier could be overcome not only thermally but also at low temperature under suitable light irradiation (see Sect. 7.3). This analogy encouraged the Utah group to investigate the spectral changes caused by laser irradiation on the low-energy side of the EA band related to simple Yb^{2+}–CN^{-}centers.

The results are shown in Fig. 9.9 for the EA, EL and VA. Starting with the VA, we can see that a reconfiguration from type a centers to type b centers indeed occurs. As indicated by the arrows, we can always identify two lines of each type. This correlation is expressed by the subscript i. The

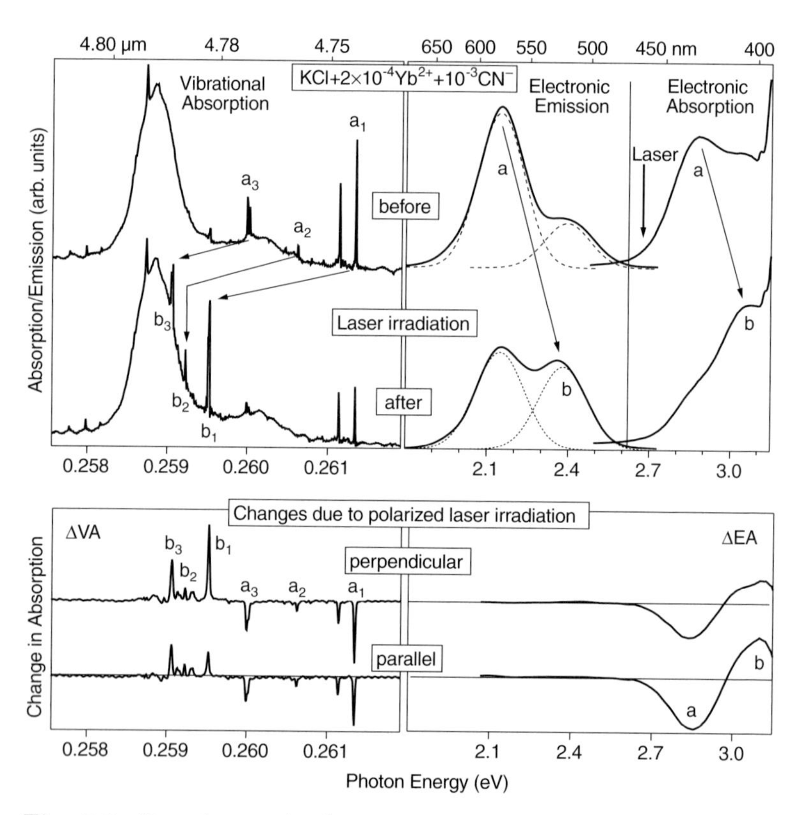

Fig. 9.9. *Top*: changes in electronic and vibrational absorption as well as electronic luminescence due to laser irradiation at 456 nm for 1 h at $T = 15$ K for a KCl:Yb^{2+}:CN^{-} sample. *Bottom*: changes in absorption induced by (100)-polarized laser irradiation

change in the VA spectra is accompanied by similar changes in the EA and EL. An unambiguous (but less detailed) assignment can be obtained which reveals that the electronic transitions move in a direction opposite to the vibrational transitions. However, both types of transitions follow the rule that a transition related to type b is shifted less than a transition related to type a, relative to the corresponding unperturbed defect (i.e. Yb^{2+} in EL and CN^- in VL respectively). *The interaction effect is clearly mutual.* These bistability experiments allowed us to decompose the electronic emission into contributions from a and b. Although the resulting spectra still maintain an inhomogeneity with respect to unresolved EA responses from a_1, a_2, and a_3 centers and from b_1, b_2, and b_3 centers, and the expected phonon structure remains obscured, the width of the bands can be used to give an upper limit for the coupling constant, $S_{\text{ph}} < 10$, which describes the interaction of the Yb^{2+} ion with regular lattice phonons.

For temperatures below 20 K, the optically produced nonthermal distribution is stable for hours but the reconfiguration can be reversed by heating the sample ($T > 100$ K) and/or by light of somewhat shorter wavelength (420 nm). While this behavior is quite similar to that of $F_H(OH)$ centers, in contrast to the latter no change in the spectral width of the lines can be observed, indicating that both types of centers in the present case are strongly orientationally aligned.

Drastic differences are found between the individual center types within the groups a_i and b_i with respect to the conversion efficiency. This is visualized in Fig. 9.10, in which, under constant laser irradiation, the buildup (for type b) and reduction (for type a) are shown as a function of time. While a_2 is changed to b_2 within a few seconds, it takes several minutes for the $a_1 \rightarrow b_1$

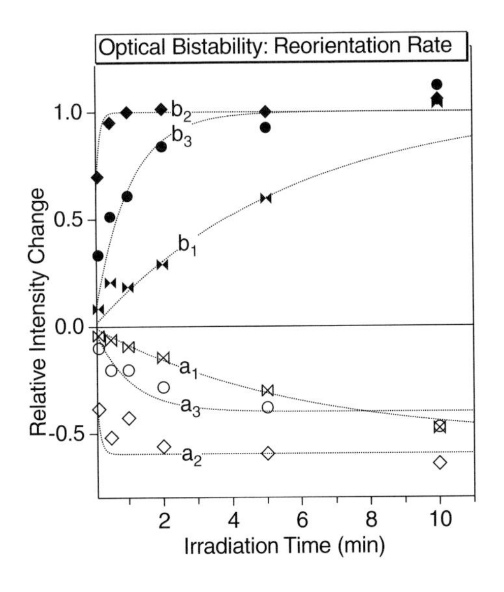

Fig. 9.10. Time evolution of the a $\rightarrow$ b conversion of Yb^{2+}–CN^- centers in KCl performed with 456 nm laser light

conversion. In any case, the efficiency is considerably lower than for $F_H(OH^-)$, with its almost optimal quantum efficiency of conversion. By comparing the VA intensity differences between the a_i and b_i centers, we find a changed transition probability of the two types of centers: the oscillator strength for b_i is in all cases almost twice that for type a. The possibility of manipulating the relative arrangement of the defect partners optically can be further exploited to learn more about the microscopic configurations of the centers, as has been demonstrated, for instance, by Luty [12] for the F_A center in alkali halides. In particular, the measurement of the linear dichroism, induced when polarized light is used in the reorientation process is very informative. However, a polarization-selective reorientation of the centers can occur only if (as for the F_A centers) a polarization of the excited states is present, a condition which was not fulfilled for the F_H centers discussed in Sect. 7.3. In contrast, Fig. 9.9 shows in both the electronic and the vibrational absorption, a $\langle 100 \rangle$ dichroic behavior in the changes of the absorption induced by a center conversion performed with (100)-polarized light. No dichroism could be produced using (110) light. Although the effect is not very strong, several clear results can be obtained:

- For the experimental conditions chosen, the degree of polarization in VA is quite different for different centers. It is highest for a_1 and b_1, and almost zero for a_2 and b_2.
- The direction of the polarization of the change in absorption is the same for the partners – best seen for a_1 and b_1. As the vibrational absorption is polarized along the molecular axis, this means that the orientation of the molecule is either changed by 180 degree or not changed at all.
- The dichroism of the EA and VA is of opposite sign which leads to the conclusion that either the molecular axis or the polarization of the lowest electronic state is perpendicular to the axis of the Yb^{2+}–CN^- pair.

Among the numerous interesting results obtained from the optical and thermal bistability experiments, we shall keep the following in mind for further discussion:

- Six simple Yb^{2+}–CN^- center types exist, of which two can always be transformed into each other by optical excitation.
- The spectral shifts in the VA and EA due to the interaction between Yb^{2+} and the CN^- ions change synchronously.
- The $\langle 100 \rangle$ axis is a preferential one within the defect pair.

9.5 Center Model

On the basis of the number of vibrational transitions and their polarization dependence (observed in the bistability measurements) and of the considerations about the origin of the spectral shifts, a model for the configuration of the centers can be deduced. On inspecting the crystal structure, one easily finds that several arrangements are possible for the simple Yb^{2+}–CN^- complexes. Assuming that the CN^- ion is in a $\langle 100 \rangle$ neighboring position of the Yb^{2+} ion, with its molecular axis along the $\langle 100 \rangle$ axis, the 12 possible $\langle 110 \rangle$ positions of the charge-compensating vacancy can be classified into three groups (i, ii, iii) with four equivalent arrangements. For each of the three groups two orientations (a, b) of the CN^- electric dipole are possible. The resulting six different configurations are shown in Fig. 9.11. In this picture, the bistable pairs a_i and b_i are distinguished by the relative orientation of the CN^- molecule. The phenomenological model for the origin of the spectral shifts, which will be discussed in Sect. 9.6, suggests that the configuration in which the carbon atom points towards the Yb^{2+} ion will yield the larger spectral shifts as observed in metal–ion complexes. The question of which group (i, ii, or iii) belongs to which VA transition (a_i, or b_i) is more difficult to answer. In the most probable assignment, the charge-compensating vacancy is closer to the CN^- molecule, the VA transitions are shifted less to the blue because both more space is available (giving a redshift) and the effective charge of the Yb^{2+} ion is reduced resulting in less charge redistribution. This leads to the assignment (iii) $\equiv$ 1, (ii) $\equiv$ 2 and (i) $\equiv$ 3.

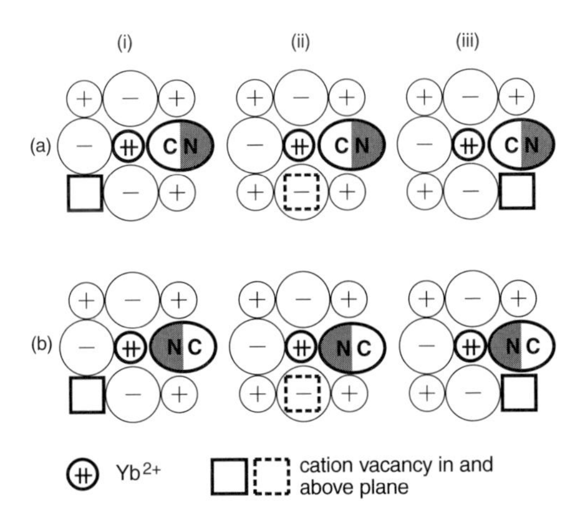

Fig. 9.11. Six possible center configurations for the Yb^{2+}–CN^- defects

In the proposed center model, the different center types have the same statistical multiplicity and it would be expected that at high temperatures they would be present in equal amounts. This is different from the case of bistable $F_H(OH^-)$ defects, for which the two center types have different entropy factors which strongly influence the thermal equilibrium. The expectation from our model is fulfilled if we extrapolate the low-temperature fit in Fig. 9.6 to high T and take into account the oscillator strength differences.

However, the extrapolated $I(\mathrm{b})/I(\mathrm{a})$ ratio for the high-temperature VA data is higher suggesting that in the present system also, the orientational alignment will be lost for one type of center increasing the entropy of that center. The other possibility of a temperature-dependence in the relative absorption cross-sections cannot be excluded, however.

Owing to the singlet GS of the centers under consideration, no standard magnetic resonance experiments are possible. Moreover, no EPR signal was found in optically detected EPR in attempts to probe the lowest excited state.

The possible number of configurations is considerably increased if more CN^- molecules are involved, leading to the observed inhomogeneous broadening of the spectra which makes a distinction impossible. The value of n can therefore be considered only as an approximate average number of neighbors. We are always dealing in these cases with a mixture of complexes with different relative center arrangements.

9.6 Interpretation of the Spectral Shifts

In this section we shall discuss the drastic shift of the Yb^{2+} transition energy caused by the CN^- neighbors. Since ab initio quantum chemical calculations of the electronic energy levels are very complicated and beyond the scope of this book, we apply here the more phenomenological approach of ligand field theory, which was developed in detail for transition-metal-ion complexes [13, 14]. This approach of treating our defect system as a single entity ("pseudo-molecule") is further motivated by the observation that similar spectral shifts of molecular Yb^{2+}:CN^- complexes are observed for $YbCN_2$ molecules in powder form [15], in alkali halide melts, and in aqueous solution. Following the usual procedure, we first determine the phenomenological ligand field strengths D_q for our complexes, and then try to explain their sizes with qualitative arguments.

9.6.1 Ligand Feld Strength

As outlined in Sect. 2.2.1, ligand field theory has been applied with varios modifications to the isolated Yb^{2+} ions in alkali halides by several authors [16, 17, 18]. Assuming octahedral (O_h) symmetry, i.e. neglecting the charge-compensating vacancy, these authors have fitted the measured absorption spectra (**A** to **H** bands) by varying the phenomenological ligand field strength parameter D_q. The value $D_q = +1240\ \mathrm{cm}^{-1}$ obtained within this model yields reasonable agreement with the absorption spectra of isolated Yb^{2+} defects in KCl [16].

A similar crystal field analysis for our Yb^{2+}:$(CN^{2+})_n$ complexes is unfortunately not possible, because the absorption spectra could not be decomposed completely into the contributions of the different complexes over a sufficiently large spectral range. For this reason only a semiquantitative comparison is possible; this has been done on the basis of published energy-level schemes, considering only the lowest-lying transition (i.e. the **A** band in absorption).

The measured energy positions of the **A** bands for the Yb^{2+}:$(CN^{2+})_n$ complexes are plotted on the left axis in Fig. 9.12 as a function of n. This measured **A** position can be converted into the ligand field strength parameter D_q with the help of the calculated energy-level diagrams (e.g. Fig. 1 of [18]). Unfortunately, the published level schemes do not extend to large enough values to allow us directly find D_q for $n = 6$ in pure KCN. For the purpose of an approximation, the lowest energy-level curve can be extrapolated linearly beyond the limit of the published figure ($D_q = 1800$ cm^{-1}), and the D_q values of our complexes are shown on the right axis of Fig. 9.12. The value of $D_q \approx 3450$ cm^{-1} obtained for the Yb^{2+}:$(CN^-)_6$ defect complex represents an increase of D_q by a factor of about 3 when all six ligands are changed from Cl^- to CN^-.

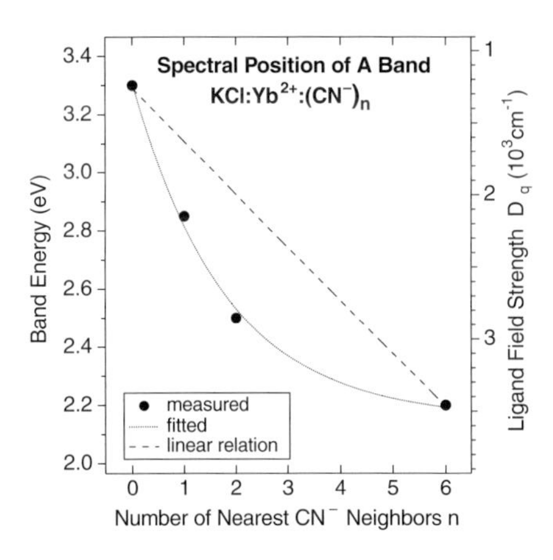

Fig. 9.12. Spectral position of A-type absorption band as a function of the number of nearest CN^- neighbors (*left axis*) and the corresponding ligand field strength D_q (*right axis*)

Phenomenologically, the experimental data (•) can be fitted to a relation (solid line in Fig. 9.12) of the following form:

$$D_q(\mathrm{cm}^{-1}) \approx 3490 - 2250 \exp(-0.6n) \,. \tag{9.3}$$

This functional behavior, with its decreasing role for every extra ligand, is commonly observed in cases where ligand–ligand interaction and covalent bonding effects play a role. It is considerably different from the linear behavior of the form (plotted as a dashed line)

$$D_q(n) \approx \frac{n}{6} D_q(6) + \left(6 - \frac{n}{6}\right) D_q(0) \,, \tag{9.4}$$

[13], expected in mixed-ligand cases in which the interaction is additive (such as a purely electrostatic interaction). As can be seen in Fig. 9.2, the increase

of ligand field strength with increasing number of CN^- defect partners is also manifested in an increase of the separation between the lowest **A**-type absorption band and the A*-type emission. As indicated in Fig. 2.3 the splitting between the lowest state T_{1u} and the E_u- and T_{2u}-type states increases with increasing cubic crystal field. The same holds for the crystal fields of lower symmetry that are encountered for our complexes.

Interpretation of the Large D_q Values. The spectral shifts and resulting D_q values for our $Yb^{2+}:(CN^-)_n$ complexes are very large compared with those observed for other ligands in a variety of host materials (e.g. alkali halides [16, 19, 3] alkaline-earth halides [21, 17, 20], and alkaline-earth sulphates [22]). This result reveals the special role of the CN^- molecule as a partner in a complex.[3] While this strong defect-partner dependence seems very unusual from the viewpoint of Yb^{2+} defects in solid-state materials, it is well known for metal-ion complexes. For such complexes involving transition metal ions and various ligands, a "spectrochemical" series has been established empirically by ordering the ligands according to their contribution to the D_q value. For the ions important to us it is found [13] that:

$$I^- < Br^- < Cl^- < OH^- \ll CN^- , \qquad (9.5)$$

and for coordinated atoms

$$I < Br < Cl < S < F < O < N < C . \qquad (9.6)$$

In these series the halide ions (Cl^-, Br^-, I^-) are on the weak-ligand-field side, whereas CN^- ligands are at the other end and always cause large D_q values. For instance, in $[CrCl_6]^{3-}$ and $[CrCN_6]^{3-}$ complexes the D_q values of 1360 cm^{-1} and 2630 cm^{-1} show a change by a factor of about two, which is comparable to the results for our $Yb^{2+}:(Cl^-)_6$ and $Yb^{2+}:(CN^-)_6$ complexes [13]. The data leading to the series have been taken mainly from d $\rightarrow$ d transitions of transition-metal-ion complexes. However, the series also can be applied for semi-quantitative purposes to the 4f $\rightarrow$ 5d transitions of Yb^{2+} owing to the small change of 4f states (i.e. the $4f$ orbitals are protected by the filled outer shells of the 5s and 5p states). Under these conditions, an increase in the splitting of the 5d levels will cause a redshift of the lowest excited states. This agreement with our experimental findings shows that the phenomenological "rule" of the spectrochemical series appears to be applicable to our system as well. For a deeper understanding of the wide variation of ligand field strengths, both purely electrostatic effects and various chemical bonding effects have to be taken into account.

[3] The only observation that is at all similar to our results is a strongly redshifted emission related to photoionization that has been found [20, 23] in several hosts, but not including the alkali halides. In these cases no shift of the absorption is observed.

In order to get a feeling for the roles of the different contributions we first estimate the spectral shifts of our systems caused by electrostatic interaction by using some equations from [24]. For that purpose, our complexes are considered as $Yb(CN)_n(Cl)_{6-n}$ molecules embedded in a lattice which restricts the possible bond lengths. In order to give a rough estimate of the required CN^- displacement ΔR, we use the point charge approximation of ligand field theory, in which the contributions to the splitting of the d-levels $(10D_q)_{\text{charge}}$ and $(10D_q)_{\text{dipole}}$ from a point charge e and a point dipole $e\Delta R$ can be written, for octahedral symmetry, as:

$$(10D_q)_{\text{charge}} = \frac{5e\overline{r^4}}{3R^5} \tag{9.7}$$

$$(10D_q)_{\text{dipole}} = \frac{25e\Delta R \cdot \overline{r^4}}{3R^6} \approx \frac{5\Delta R}{R}(10D_q)_{\text{charge}} \tag{9.8}$$

where r is the radial coordinate of the 5d electrons of the central ion, $\overline{r^4}$ is the expectation value of r^4 and R is the distance between the central ion and the ligands. If , as a test, we make the crude assumption that the D_q value for our Yb^{2+} defects in KCl is due to point charges, while the change in KCN relative to KCl (i.e. $D_q(\text{KCN}) \approx 3D_q(\text{KCl})$) is caused by the displacement dipole, we find a value for the displacement of $\Delta R \approx \frac{2}{5} \cdot R$. This is an unreasonably large off-center shift of the CN^- ion. Similarly, the permanent electric dipole of the CN^- ion is also too small to account for the large D_q values. For these reasons, the electrostatic contribution can be ruled out as the sole source of the increased D_q values. Obviously, within our complexes metal–ligand exchange effects [25], covalency, and/or charge transfer [26] become appreciable.

As a second step, we consider a charge transfer or redistribution from the CN^- molecule to the Yb^{2+} ion, thereby adding a covalent contribution to the otherwise strongly ionic bond between the defect partners. Since, firstly, the electron affinity of Yb^{2+} is much larger than the ionization energies of CN^- and Cl^- and, secondly, the Yb^{2+} ion represents an effective positive charge within the alkali halide lattice, we expect that the redistribution of electronic charge density will be directed towards the Yb^{2+} ion leading to a decrease of its effective ionization state. To estimate the influence of this effect, it is helpful to consider, as shown in Fig. 9.13, the effect of decreasing the ionization of Yb in the series of free ions and atoms $Yb^{2+} \rightarrow Yb^+ \rightarrow Yb^0$.

As the ionization number decreases ($m = 2 \rightarrow 1 \rightarrow 0$) and electrons are added to the $6s$ orbital the energy of the 5d orbitals is lowered and consequently the energy of the 4f $\rightarrow$ 5d transition (Fig. 9.13). However, even for neutral Yb^0 this change is smaller than that observed for our Yb^{2+}:$(CN^-)_n$ complexes in KCN, so that we have to look into even finer details.

In the chemistry of metal ion complexes, the covalent contribution to a bond is often discussed in terms of the "nephelauxetic" series, which describes the expansion of the electron cloud of the central ion due to the

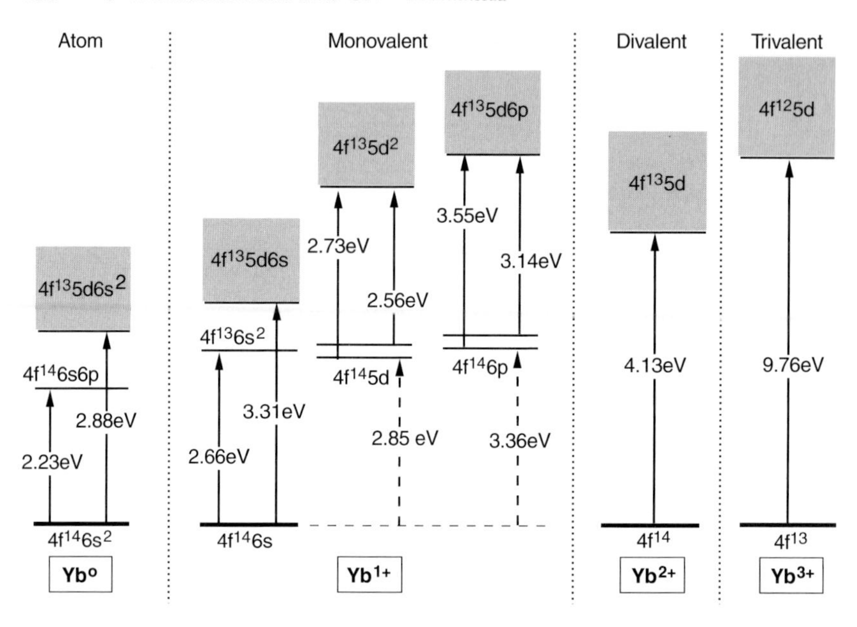

Fig. 9.13. Energy levels and possible optical transitions for Yb in different charge states. For Yb^{1+}, the dependence on the occupied electronic orbital is depicted

neighboring ligands [14] and is used as a measure of the covalency of the bond with the ligand. In this series (which should also hold for the 5d levels of rare-earth ions), the halide anions (Cl^-, Br^-) are located very close to the CN^- ion giving another apparent argument against our attempt to explain the spectral shifts by covalency. However, we want to argue that not the degree of the bonding but the type of orbitals involved in the bond determines the transition energies. This assumption is motivated by Mössbauer spectral measurements: In thoses studies it was found, for $Eu(CN)_2$ compared with $Eu(Cl)_2$, that while the bonding is only slightly more covalent, the 5d orbitals of Eu^{2+} participate much more in the bonding [27]. Although such data are not available for Yb^{2+}, it is reasonable to assume a similar behavior owing to the electronic similarities. Indeed, the general behavior of Yb^{2+}:$(CN^-)_n$ and Eu^{2+}:$(CN^-)_n$ defect complexes is very similar, as will be shown in Chap. 10. In order to follow up this idea further, we compare in Fig. 9.13 how the transition energy is altered when the extra electron in the case of Yb^1 is added not to the 6s but to the 5d orbital. As can be seen, a drastic change is found, giving strong support for an interpretation in which the observed large changes in D_q are due to a redistribution of charge from the CN^- ligand to the 5d orbital of the Yb, whereas for Cl^- and Br^- the charge is transferred to the 6s and 6p orbitals. This difference in the *nature* of the bonding can also explain the different relative positions of CN^- in the nephelauxetic and spectrochemical series.

Vibrational Transitions. Encouraged by this plausible explanation of the changes in the Yb^{2+} transitions and taking advantage of the "second side" of our defect system, we shall now test this simple picture in an attempt to explain the changes in CN^- vibrational transitions.

As reviewed in Sect. 3.1.2, the energy shifts of the vibrational transitions of a molecule due to its interaction with the ions within a crystal are usually dominated by a repulsive interaction, leading to a decrease in the eigenfrequency as the space available for the molecule increases. Therefore, in our case, a redshift of the CN^- transitions would be expected at first sight because the Yb^{2+} ion and the required charge-compensating cation vacancy are certainly smaller than the replaced alkali ions. However, the opposite is observed. This must be caused by the charge redistribution effect discussed above. In order to elucidate this matter, more details of the intramolecular CN^- bonding have to be considered. The molecular orbitals of the CN^- ions are shown in Fig. 9.14.

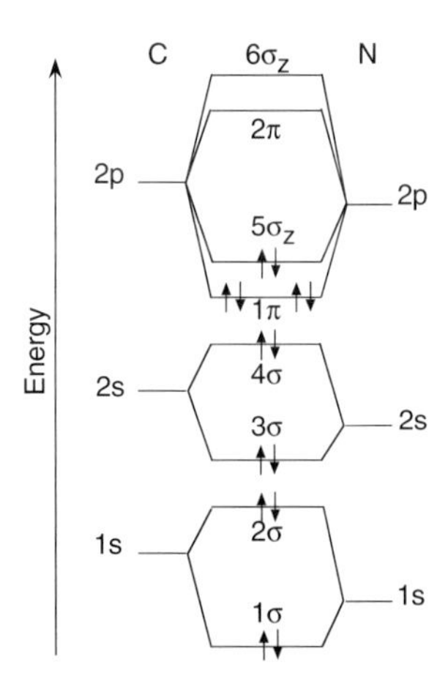

Fig. 9.14. Schematic picture of molecular orbitals involved in the bonding of a CN^- molecule

In several metal complexes (e.g. AgCN and HgCN) and organic molecules with CN^- constituents, an increased eigenfrequency has been found, especially in those cases in which the molecule forms a σ-type bond with another partner. The formation of such a metal-ion/molecule bond shifts one of the σ_z electrons out of the intramolecular CN^- triple bond. This is exactly what happens in our complexes as well. The double positively charged Yb^{2+} ion attracts the electron of the CN^- molecule, causing a charge redistribution with an average shift out of the CN^- intramolecular bond towards the Yb^{2+} ion. This leads to a stiffening of the CN^- bond and an increase of the vibrational eigenfrequency. The charge redistribution can be considered as the formation of a covalent bond between the CN^- and the Yb^{2+} ions which is, however, not very strong owing to the restriction on the positions of the defect partners caused by the lattice. For that reason, the energy shift of the CN^-

ion is less than what has been observed for "true" $YbCN_2$ molecules [15]. To summarize this subsection, let us recall the essence of our phenomenological interpretation:

- Energy shifts are caused by a charge redistribution from the CN^- molecule to the Yb^{2+} ion.
- The kind of Yb^{2+} electronic orbital into which the charge is transferred is of critical importance. Involvement of the 5d orbital leads to much stronger spectral shifts.

For a more quantitative understanding, a quantum chemical treatment of the molecular structures of the Yb^{2+}:$(CN^-)_n$ complexes is necessary, taking both the charge distribution among the defect partners and the influence of the lattice into account. Application of the recently developed method of embedded-molecular-cluster calculations (which has been successfully used for $[Fe(CN^-)_6]^{3-}$ in NaCl [28]) to our system appears to be most promising [29]. The experimental and theoretical work on the $[Fe(CN^-)_6]^{3-}$ complex was sparked by commercial interest, since that complex is expected to play an important role in the formation of latent images in photographic films. The results reveal several similarities to our defect system. Unfortunately, the analogous work on Yb^{2+}–CN^- complexes has not been completed yet, but a comparison with the results for the $[Fe(CN^-)_6]^{3-}$ complexes gives many indications that our experimental results can be reproduced and that further information about the defect structures and the electronic configuration can be extracted.

9.7 Vibrational Luminescence and E–V Energy Transfer

Turning our attention from the static shifts in the transition energies of the defect partners to the dynamic properties of the defects and the changes occurring during the excitation–relaxation cycle, we shall treat now the E–V energy transfer and the occurrence of vibrational luminescence. The possibility of observing CN^- vibrational luminescence, which has a long radiative lifetime in most alkali halide host materials, gives a fairly simple and powerful tool to investigate the E–V energy transfer. Such an investigation will be described in the following, starting with the cw measurements of the VL spectra and their interpretation.

Examples of VL spectra are shown in Fig. 9.15. The vibrational luminescence of Yb^{2+}:$(CN^-)_n$ complexes can be observed using optical excitation with an Ar^+ laser ($\lambda = 457$ nm, 488 nm, 515 nm). For low CN^- concentrations (Fig. 9.15a, bottom) at least seven equidistant ($\Delta \approx 25$ cm^{-1}) lines are observed, which are characteristic of a sequence of anharmonicity-shifted transitions from different excited v-levels. Their spectral positions are different from those expected for isolated CN^- defects, and the lines are therefore assigned to CN^- defects perturbed by Yb^{2+} neighbors. The corresponding transitions are marked, showing that vibrational levels up to $v = 8$ can be excited. This assignment is supported by the observation that when the isotopic composition of the CN^- ion is varied from $^{12}C^{14}N^-$ to $^{13}C^{14}N^-$, the sequence of lines shifts by 44 cm^{-1} to lower energies owing to the higher reduced mass of the latter isotopic species. Besides this, it can be seen in Fig. 9.15 that for higher CN^- concentrations VL from isolated CN^- defects also appears; these defects are obviously populated via V–V transfer [30].

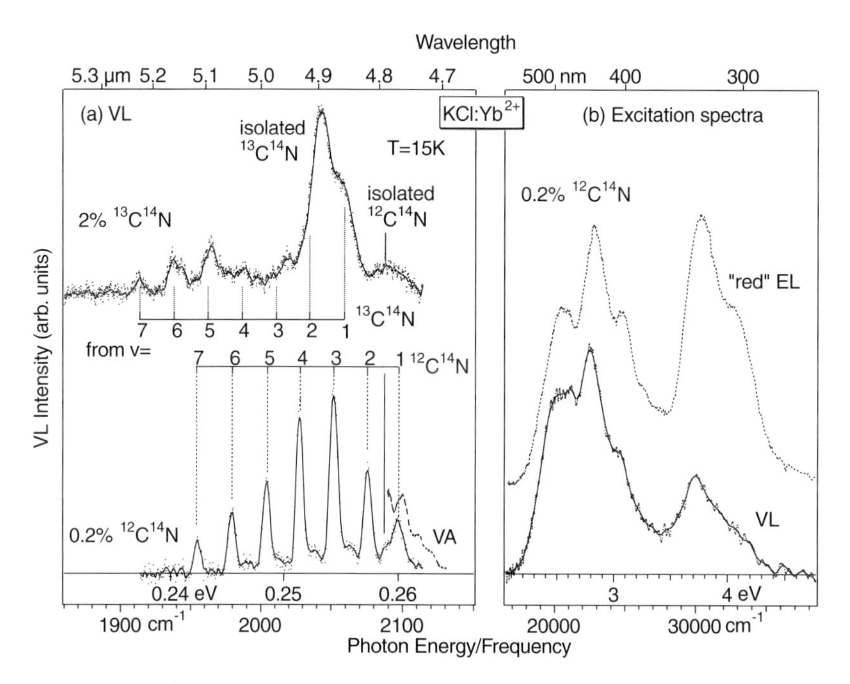

Fig. 9.15. (**a**) Vibrational luminescence under $\lambda = 514$ nm Ar^+ laser excitation in KCl doped with $4\times10^{-4}Yb^{2+}$ and different CN^- isotopic species and doping levels: *top*, 2% $^{13}C^{14}N^-$; *bottom*, 0.2% $^{12}C^{14}N$. The starting levels of the anharmonicity-shifted transitions of the Yb^{2+}:$(CN^-)_n$ defects are indicated, as well as the $1\to0$ transition of the isolated defect. (**b**) VL excitation spectrum comparised with that for the red luminescence

The VL results for RbCl and KBr are quite similar to those presented for KCl. A closer look at the results for the NaCl host material is more interesting, firstly because the vibrational decay of isolated defects in this host is strongly nonradiative owing to an effective coupling to local modes [31, 32], and secondly because, as we have seen above, there is a strong tendency towards defect clustering and a significant increase in the VA cross section.

Quite differently from thc F center defect systems in this host material, it was possible to observe a fairly strong VL luminescence signal, suggesting a high E–V transfer efficiency (see Fig. 9.16). Similarly to the situation for

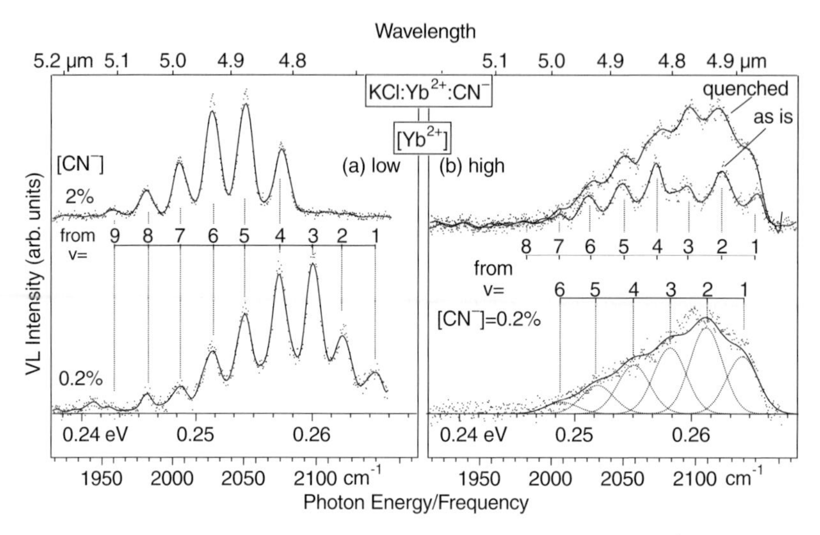

Fig. 9.16. Vibrational luminescence under λ = 488 nm Ar^+ laser excitation in NaCl. (**a**) Low [Yb^{2+}] concentration ($\approx 2 \times 10^{-4}$) for two different CN^- concentrations. (**b**) High Yb^{2+} concentration ($\approx 2 \times 10^{-3}$) before and after quenching of the samples (5 min, 120 °C). At the *bottom*, the difference is fitted to six anharmonicity-shifted Gaussian curves. See text for details. The assignments of the different peaks are indicated by the *dashed lines*

KCl, the observed equidistant lines can be assigned to anharmonicity-shifted transitions from excited v-levels using the peak with high energy as the starting point; the corresponding value of $\overline{\omega}_{10}$ is included in Table 9.1. It can be seen that fairly high vibrational states (up to $v \approx 9$) are populated again. Comparing the spectra for different CN^- concentrations, it is found for high [CN^-] that the lower vibrational transitions are missing, most likely owing to V–V energy transfer to isolated defects. However, no VL is observed from the latter, suggesting that the contribution of nonradiative decay channels is more pronounced for the isolated CN^- molecules than for those perturbed by Yb^{2+} ions.

In VA spectra of highly Yb^{2+}-doped samples, it was observed that a strong broad band appears, which can be removed almost completely by thermal treatment of the sample. A similar effect is observed in VL as well. Figure 9.16b shows the VL spectra for the same crystal as was used for Fig. 9.8. If the sample was stored for several days at room temperature, we found a broad, almost unstructured band in the VL. Heating the sample to 120 °C followed by fast quenching to low temperature reduced the overall VL strength and changed the spectrum into several distinct peaks which were similar to those found for low Yb^{2+} concentrations. The reduction of VL strength could be reproduced in several runs, yielding a VL at least 10 times stronger for the unquenched samples. This suggests, similarly to the

VA results, that the center complexes responsible for the broad response have higher vibrational transition probabilities. A more detailed analysis of the VL structure is shown at the bottom of Fig. 9.16b. We have assumed here that the broad, almost Gaussian-shaped VA band (see fit in Fig. 9.8) appears as the $v = 1 \rightarrow 0$ transition in VL. The transitions from higher vibrational levels are obtained by shifting the curve according to the anharmonicity by consecutive steps of $\Delta\omega = 25\ \mathrm{cm}^{-1}$. In order to take the contribution from the defect centers with sharper VL spectra into account, we subtracted from spectrum (1) (see Fig. 9.16b) a fraction (≈ 0.65) of spectrum (2) so that no structure on top of the broad band remained. A least-squares fit of the six Gaussian curves to the measured data yields fairly good agreement. Comparing the vibrational transitions obtained for high and low Yb^{2+} concentrations, for samples having similar CN^- concentrations, it is found that while the spectral positions are slightly different, the distribution of relative strength is almost identical. Obviously, the energy transfer and relaxation processes for complexes involving just a single Yb^{2+} ion and those with more Yb^{2+} ions are similar.

9.7.1 Type of Center Involved in the E–V Energy Transfer

The important question of which type of defect takes part in the E–V energy transfer, can be answered in several ways:

- The spectral positions of the VL and VA can be compared. It can be seen clearly that none of the VA lines of the simple Yb^{2+}–CN^{2+} pair coincides with any of the VL lines, either in position or in spectral width. This rules out the simple defect type as the main contributor to the E–V energy transfer. On the other hand, a good coincidence of the spectral properties for the centers involving more CN^- molecules is found.
- The obvious conclusion that only these more complex centers participate in the E–V transfer is further supported by the excitation spectrum for the VL (shown in Fig. 9.15b) which resembles that for the red luminescence ($n \geq 2$) very closely and is characteristically different from the spectrum of the orange luminescence. A closer examination of the spectra, particular in those spectral regions in which no absorption by centers with $n \geq 2$ exists, shows that less than 10% of the VL can originate from simple centers with $n = 1$.
- The VL spectra are not changed if the excitation energy is chosen to be in a spectral range for which only centers with $n \geq 2$ absorb.
- A final confirmation of the result can be obtained by time dependent measurements. These measurements show (Fig. 9.17) the long ($\approx$3 ms) lifetime of the blue EL from isolated Yb^{2+}, which is reduced only slightly for the Yb^{2+}–CN^- complex (orange EL), suggesting that little or no E–V transfer occurs. On the other hand, the lifetime of the red EL (due to Yb^{2+}:$(CN^-)_{n=2,\ldots}$) is reduced significantly (2 ms). Moreover, the decay

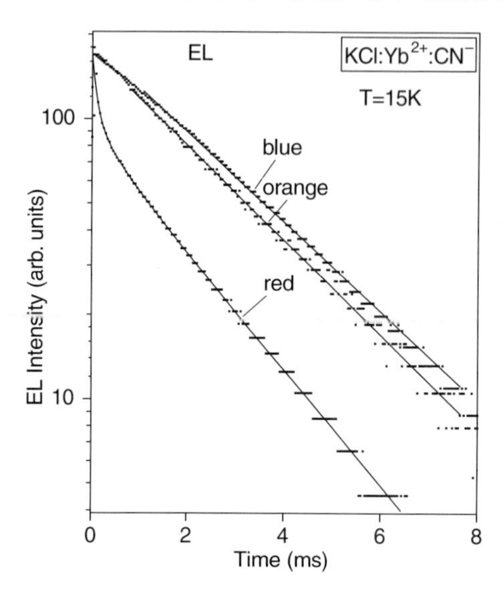

Fig. 9.17. Decay curves (on a log scale) of the EL after 40 ns pulses from an excimer dye laser with an appropriately selected wavelength (420 nm for blue, 520 nm for orange, 580 nm for red). The *solid lines* represent the fitted (multi-) exponential decay curves with $\tau_{\mathrm{blue}}^{-1} = 365\ \mathrm{s}^{-1}$ for the blue EL, $\tau_{\mathrm{orange}}^{-1} = 390\ \mathrm{s}^{-1}$ for the orange EL, and $\tau_{\mathrm{red}}^{-1} = 500\ \mathrm{s}^{-1}$, $1800\ \mathrm{s}^{-1}$, and $14000\ \mathrm{s}^{-1}$ for the red EL

becomes multiexponential with at least three components. For the low CN$^-$ concentration used the "slow" 2 ms component is still the dominant one and is therefore assigned to the complexes with two CN$^-$ ions. The other components are due to centers with even more CN$^-$ ions.

The consistency of the results obtained by different methods gives a nice example of how this defect system can be characterized in detail. The data presented above can be exploited further to estimate the E–V transfer efficiency. Again, several possible methods can be employed. Firstly, if the increase in the decayrate of the red EL, $\Delta k = k_{\mathrm{red}} - k_{\mathrm{orange}}$, is attributed exclusively to the additional E–V transfer the latter should have the rate $\Delta k = k_{\mathrm{e-v}} \approx 170\ \mathrm{s}^{-1}$. This means that the E–V transfer has a relative efficiency of about 30%. This result and the underlying assumption can be confirmed by measurement of the integrated photon–number for the red EL and VL, which also shows that about 30% of the emitted photons appear as VL.

An even more detailed picture for which center types with $n \geq 2$ participate in the transfer and of how they do so, can be obtained from further time dependent measurements of the dynamics of the transfer.

9.8 Dynamics of the E–V Transfer

As has been mentioned before, the determination of the time dependence of the E–V transfer proved to be a formidable task for the F center systems owing to the short transfer times involved [33, 34]. The situation is much more favorable for the Yb^{2+}:$(\mathrm{CN}^-)_n$ defect system, for which both the EL and the

VL can easily be detected within a convenient time window (milliseconds) that is known from studies of isolated CN^- and Yb^{2+} defects [31, 35]. Using the results presented above we have a well-defined parameter-rich system to hand which is very suitable for use as a new model system to investigate E–V energy transfer in great detail. The main questions that we want to address in this section are the following:

- In which part of the optical-excitation/relaxation cycle of the Yb^{2+} ion does the E–V transfer take place?
- Into which vibrational CN^- level(s) does the primary transfer occur?
- How does the energy transfer efficiency change with temperature?

In order to guide our treatment of the dynamic properties let us first introduce, on the basis of the spectroscopic results, a "working" model (see Fig. 9.18) of the excitation–relaxation cycle and the underlying processes (excitation, emission, and energy transfer). Excitation in the near-UV spectral region excites (to some extent) all the different types of Yb^{2+}-related centers (isolated Yb^{2+}, $Yb^{2+}-CN^-$, $Yb^{2+}:(CN^-)_n$, etc.) which coexist statistically in the codoped samples. After internal nonradiative relaxation, a Stokes-shifted electronic emission occurs from the lowest excited state with a wavelength characteristic of the respective center type ("blue", "orange", or "red"). Competing with this radiative decay, several energy transfer processes can take place. Firstly, for a high Yb^{2+} concentration, the excitation energy can be transferred from one Yb^{2+} center to another, either resonantly among

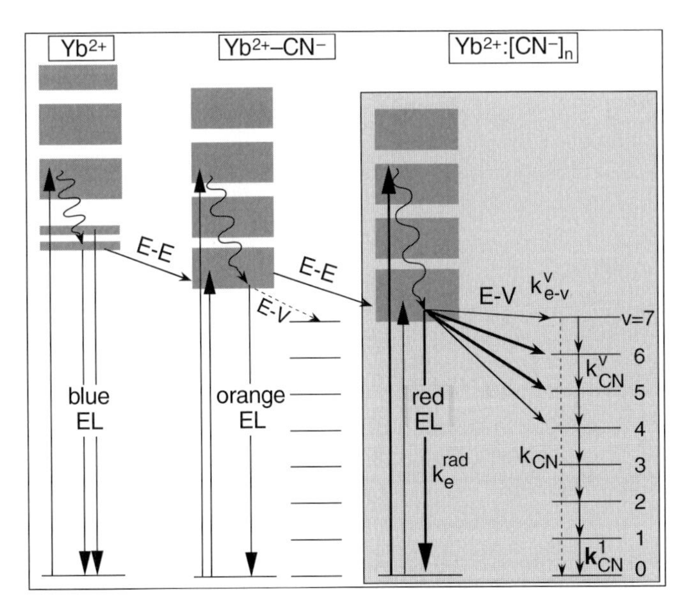

Fig. 9.18. Schematic model of the emission and energy transfer processes occurring after optical excitation of $Yb^{2+}:(CN^-)_n$ complexes (see text for details)

defects of the same type or nonresonantly, assisted by phonons, to centers with more CN^- neighbors. This E–E energy transfer will not be discussed here, as it can be avoided by using samples with a low Yb^{2+} concentration. Secondly, the presence of a strong VL shows that an efficient E–V energy transfer takes place predominantly within defects with more than one CN^- neighbor. These considerations and the scheme on the right-hand side of Fig. 9.18 can easily be converted into a dynamic model.

We obtain an initial, very crude model if we assume direct excitation of a complex (with $n > 1$) and ignore the individual vibrational levels, thereby pretending that the CN^- molecule relaxes (dotted arrow) in a single step with a rate k_{CN}. With these simplifying assumptions, the dynamics of the E–V transfer process can be described by the following rate equations:

$$\frac{dN_e}{dt} = -\left(k^e_{rad} + k_{e-v}\right) N_e \,, \tag{9.9}$$

$$\frac{dN_{CN}}{dt} = -k_{CN} N_{CN} + k_{e-v} N_e \,, \tag{9.10}$$

where N_e and N_{CN} are the numbers of centers in the electronic and vibrational excited states, respectively, and k^e_{rad}, k_{e-v}, and k_{CN} are the rates for radiative electronic relaxation, E–V transfer and vibrational relaxation, respectively. This simple system of differential equations can be solved, and we obtain, writing $k_e = k^e_{rad} + k_{e-v}$,

$$N_e(t) = N_e(0) \cdot e^{-k_e \cdot t} \,, \tag{9.11}$$

$$N_{CN}(t) = N_e(0) \cdot \frac{k_{e-v}}{k_e - k_{CN}} \left(e^{-k_{CN} \cdot t} - e^{-k_e \cdot t}\right) . \tag{9.12}$$

This solution predicts an exponential decay for the electronic luminescence and an "up–down" response for the vibrational luminescence. The solution for N_{CN} is symmetric in the two rates involved k_{CN}, k_e so that only their relative sizes determine which one (i.e. the faster one) represents the rise and which one (i. e. the slower one) represents the fall of the VL response curve. In our Yb^{2+}:$(CN^-)_n$ systems the two rates are fairly similar, and we expect to see both cases $k_{CN} > k_e$ and $k_{CN} < k_e$, depending on the host material, CN^- concentration and temperature. We have assumed in our model that the EL and the E–V transfer are competing processes starting from a common state. Only in that case one could expect to "see" the electronic decay rate directly in the time dependence of the VL. For that reason, a fit of the VL response using the EL decay rate is a crucial test for deciding where the E–V transfer takes place.

The experimental results which correspond best to the simple model described above are obtained from spectrally unresolved measurements of the red luminescence and of the "total" VL. These results are shown in Fig. 9.19 for the KCl and NaCl host materials.

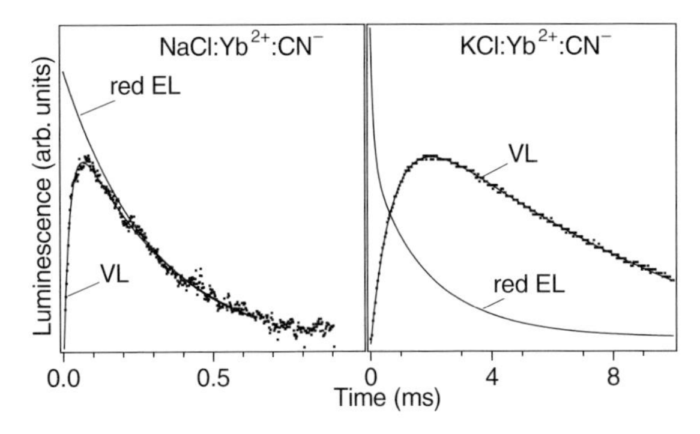

Fig. 9.19. Response curves of the VL (dots) after pulsed excitation for NaCl and KCl. The *solid lines* represent fits to (9.12) using the decay rate of the red EL

In all cases a very good fit to (9.12) (solid lines) can be obtained as long as the multiexponential decay of the red EL is included. A fit using the orange EL does not give good agreement. Comparing the results for the two hosts, one can see that indeed the two different predicted types of behavior are realized:

- In NaCl, the vibrational lifetime of the CN^- ion is shorter [32] than that for the electronic states and therefore the decay of the VL time response occurs at the same rate as the red EL. The initial increase of the VL reflects the effective vibrational CN^- decay rate which is found to be a little smaller than that observed for the isolated CN^- ion at the same temperature.
- In KCl and RbCl, the situation is exactly the opposite: the vibrational decay is slower than the electronic decay. For that reason, the initial VL increase is related (via the E–V transfer) to the EL decay, and the long decrease to the subsequent CN^- relaxation.

Apart from these differences the dynamic behavior is quite similar in the three hosts. This justifies our decision to continue to concentrate in the following on KCl, which was studied in the most detail. The experiments performed with other host materials (NaCl, RbCl, and KBr) yielded qualitatively similar results. In all cases, the good agreement with the model shows that the EL and the E–V transfer are competing processes which take place from a common electronic state.

Although this initial model proved to be very useful, it is of course too simple in several respects. In order to interpret the VL measurements of the individual VL transitions, we have to treat the multistep vibrational relaxation in more detail. After the E–V transfer into some vibrational level, the vibrational energy will relax successively down the vibrational ladder, predominantly in steps of $\Delta v = 1$. This adds more equations to our system of differential equations. These have the form

$$\frac{\mathrm{d}N_{\mathrm{CN}}^{v}}{\mathrm{d}t} = -k_{\mathrm{CN}}^{v} \cdot N_{\mathrm{CN}}^{v} + k_{\mathrm{CN}}^{v+1} \cdot N_{\mathrm{CN}}^{v+1} + k_{\mathrm{e-v}}^{v} \cdot N_{\mathrm{e}}\,, \tag{9.13}$$

where k_{CN}^v and $k_{\mathrm{e-v}}^v$ are the v-level-dependent rates of the vibrational transitions and of the E–V transfer. A formal solution can be obtained as follows:

$$N_{\mathrm{CN}}^v = \mathrm{e}^{-k_{\mathrm{CN}}^v t} \int_0^t \left[\mathrm{e}^{-k_{\mathrm{CN}}^v \tau} \left\{ k_{\mathrm{e-v}}^v N_{\mathrm{e}}(\tau) + k_{\mathrm{CN}}^{v+1} N_{\mathrm{CN}}^{v+1}(\tau) \right\} \mathrm{d}\tau \right] \tag{9.14}$$

$$= \mathrm{e}^{-k_{\mathrm{CN}}^v t} \int_0^t \left[\mathrm{e}^{-k_{\mathrm{CN}}^v \tau} \left\{ a\mathrm{EL}(\tau) + b\mathrm{VL}_{v+1}(\tau) \right\} \mathrm{d}\tau \right] . \tag{9.15}$$

The first form can be used to obtain the analytical solutions step by step, starting from the highest populated v-levels. In our case where $v_{\mathrm{max}} = 7$, the solution becomes very complicated, especially towards the lower vibrational levels. Solutions for a four-level system are given in [36]. For the fitting of our experimental $\mathrm{VL}_v(T)$ curves, the second form is more valuable; this uses the measured time response curves of the red emission $\mathrm{EL}(\tau)$ and the vibrational luminescence $\mathrm{VL}_{v+1}(\tau)$ from the level above. This allows a successive fitting of the measurements starting with the highest v-level. The parameters a and b allow one to adjust for the experimental differences in the measurements of the EL and VL.

It is interesting to note that for a harmonic oscillator, the v-dependence of the vibrational transition rates has the simple form

$$k_{\mathrm{CN}}^v = v k_{\mathrm{CN}}^1 . \tag{9.16}$$

Under these conditions, one can show that the total vibrational luminescence $\mathrm{VL}^{\mathrm{tot}}$ can be calculated as

$$\mathrm{VL}^{\mathrm{tot}}(t) = \sum_{v=1}^{\infty} k_{\mathrm{CN}}^v N_{\mathrm{CN}}^v \propto \frac{k_{\mathrm{e-v}}}{k_{\mathrm{e}} - k_{\mathrm{CN}}^1} \left(\mathrm{e}^{-k_{\mathrm{CN}}^1 \cdot t} - \mathrm{e}^{-k_{\mathrm{e}} \cdot t} \right) . \tag{9.17}$$

This is of the same form as (9.12) when the rate k_{CN} is interpreted as the decay rate k_{CN}^1 of the lowest level. This justifies the use of the simplified model above where we considered the spectrally unresolved VL response. For simplicity, we have not included the V–V energy transfer of the vibrational excitation out of the Yb^{2+}:$(\mathrm{CN}^-)_n$ complexes into more abundant isolated molecules, which is expected to play a role similar to that in the case of the $\mathrm{F_H(CN^-)}$ center, mainly for lower vibrational levels ($v = 2, 1$) and high CN^- concentration [36].

Owing to the much lower emission signal, spectrally resolved, time dependent measurements are possible only in favorable cases. For instance, we have seen earlier that a study of this type was not possible for $\mathrm{F_H(CN^-)}$ centers. We have had more luck with Yb^{2+}:$(\mathrm{CN}^-)_n$ complexes (Fig. 9.20). The VL from the higher v-states peaks before the VL from the lower states, suggesting that the model developed above is applicable, i.e. the E–V transfer occurs directly into the upper levels. Population of the higher vibrational levels by multiple excitation [37], as seen for $\mathrm{F_H(OH)^-}$ defects for example (see Sect. 7.4), or by V–V transfer [30] does not play a role. This important result is also confirmed by the observation that the decay curves are independent of the excitation power. Using the formal solution (9.15), the experimental curves (dots) can be fitted (solid lines), showing a fairly good

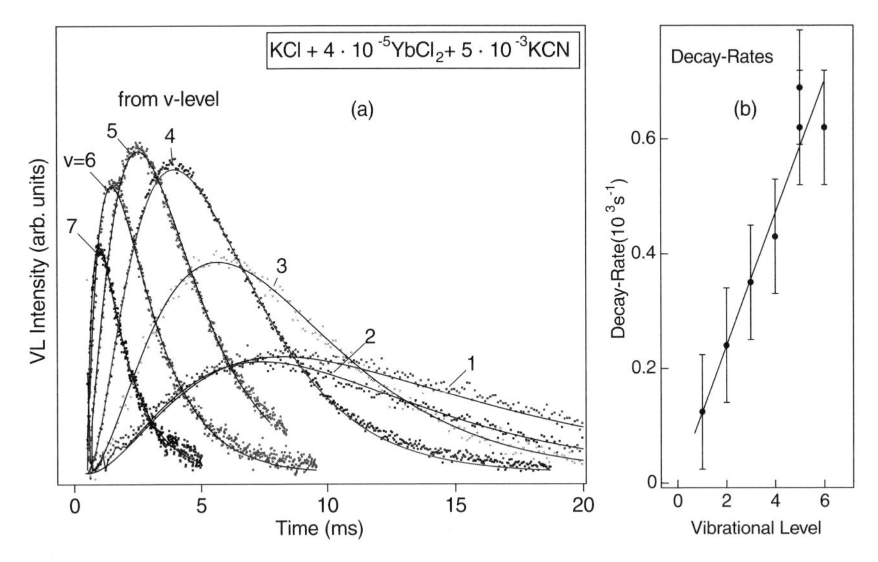

Fig. 9.20. (**a**) Spectrally resolved time response of VL (*dots*) from individual vibrational levels. The *solid lines* represent fits using (9.15) and the decay rate of the red emission. The v-dependent vibrational decay rates (determined as fit parameters) are shown in (**b**). The *solid line* shows a fit to a linear v-dependence

agreement. In order to obtain a good fit for the lower levels $v = 1, 2$, V–V transfer to isolated molecules has to be taken into account. From the fit, the v-level-dependent CN^- decay rates and relative E–V transfer rates can be obtained. The resulting decay rates are shown in Fig. 9.20b and the transfer rates are included as black bars in Fig. 9.29, which will be discussed later on.

The CN^- decay rates increase with increasing v almost linearly (indicated by the line in Fig. 9.20b) as expected for a harmonic oscillator. The rate $k^1_{\mathrm{CN}} \approx 100\ \mathrm{s}^{-1}$ obtained from the slope is about equal to the rate determined from the total VL curves (Fig. 9.19). However, the rate is higher than the radiative rate expected from the absorption strength, reflecting the role of nonradiative processes. The measurements of the individual levels and the fits to the dynamic model are fairly tedious and hence have only been done for the simplest case of low temperature and low CN^- concentration. For that reason we shall limit ourselves in the following to the existing measurements of the total VL, which still give (as shown) valuable information about the dynamics of the system.

9.8.1 Temperature Dependence

As the first variation away from the simplest case of very low temperature and low CN^- concentration, Fig. 9.21 shows the time response of the total VL

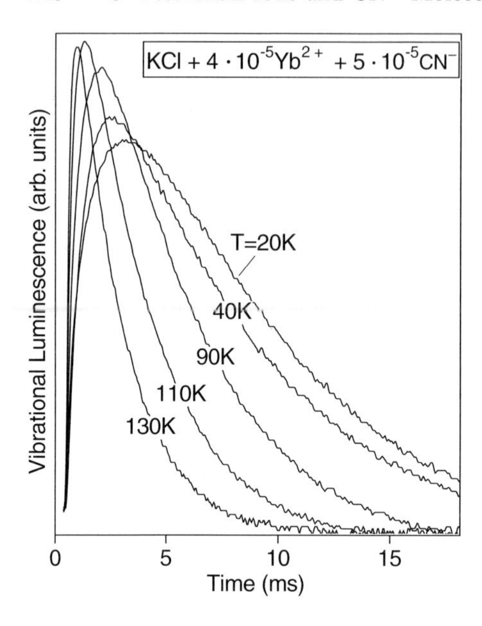

Fig. 9.21. Time dependence of VL after pulsed excitation of the total (not resolved spectrally) vibrational luminescence for various temperatures

as a function of temperature. The curves change considerably exhibiting an increased rate of both rise and decay, but the VL maxima reached are almost constant. Using again the simplified v-level-independent model to evaluate the curves, the decay rates of the red EL and VL can be determined and are shown in Fig. 9.22.

As expected from the model, the rates for the red EL obtained directly from EL measurements and from the total VL are identical. In Fig. 9.22a those rates are shown in comparison with the rates for the orange and blue EL. As a common feature, one can observe that all three EL decay rates increase with temperature. This tendency becomes most pronounced for defects with several CN^- partners. One might suspect at first that the reduced lifetimes are caused by an increase of the nonradiative processes. However, the measurement of the integrated red EL, shown in Fig. 9.22b, does not exhibit a decrease, but rather exhibits a small increase with temperature. For that reason, one has to conclude that the radiative transition rate is increasing. The VL shows the "normal" behavior well known already from studies of isolated CN^- ions: the CN^- vibrational lifetime (see Fig. 9.22a) and the integrated VL strength (Fig. 9.22b) decrease with T owing to an increase in the competing nonradiative decay channels. In order to determine the temperature dependence of the E–V transfer efficiency, one has to correct for these losses. The important result is shown in Fig. 9.22c: *although the radiative decay rate of the red* EL *varies by two orders of magnitude, the* E–V *energy transfer efficiency stays constant and hence the* E–V *transfer rates* $k_{\mathrm{e-v}}$ *are almost proportional to the radiative decay rates* $k_{\mathrm{e}}^{\mathrm{rad}}$.

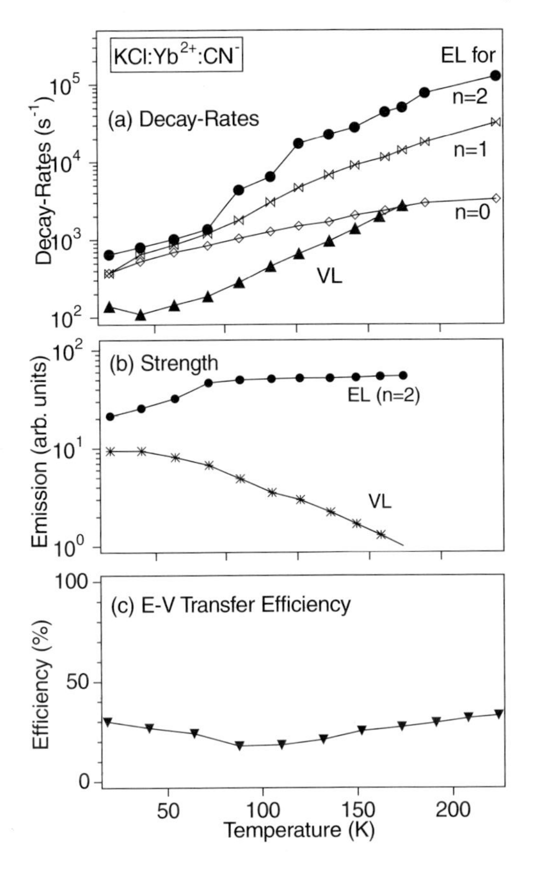

Fig. 9.22. Temperature dependence of (**a**) the decay rate of VL andof the EL from different types of Yb^{2+}:$(CN^-)_n$ centers, (**b**) the integrated strength of the EL and VL, and (**c**) the E–V transfer efficiency obtained using the VL decay rate shown in (a) and the EL and VL strengths

9.8.2 Concentration Variation

We have already seen for low CN^- concentrations that the red emission originates from various center types belonging to the family of the complexes Yb^{2+}:$(CN^-)_n$. This inhomogeneity becomes even more pronounced for higher CN^- concentrations. In such samples a wide range of complexes with various numbers n of CN^- neighbors is present. As was shown in Sect. 9.2, the EL peak positions shift into the IR for higher-doped samples, especially if the excitation energy is decreased, because Yb^{2+}:$(CN^-)_n$ centers with more and more CN^- partners are excited. The question of to what extent the E–V energy transfer is influenced by this increase in the number of CN^- neighbors and the associated decrease in energy of the relaxed excited state was studied in various ways as described below.

E–V Transfer Efficiency Spectrum. In one sample with a high CN^- concentration the relative E–V transfer efficiencies, $\eta_{\mathrm{e-v}} = \frac{\mathrm{VL}}{\mathrm{VL+EL}}$, were determined by simultaneously detecting the EL and VL while the excitation energy was varied (Fig. 9.23). This efficiency shows a significant increase to-

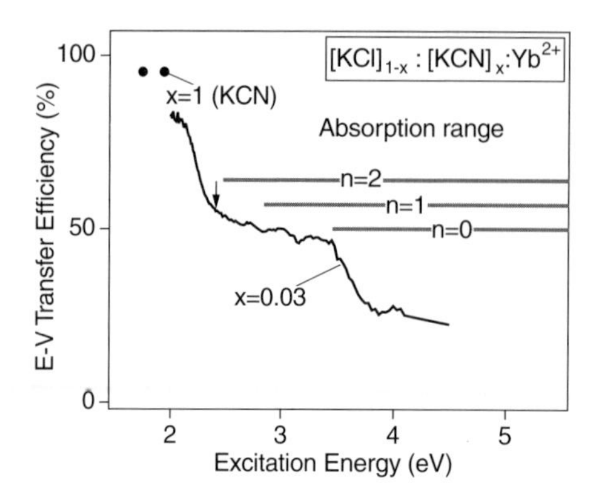

Fig. 9.23. Relative E–V transfer efficiency in the $(KCl)_{1-x}$:$(KCN)_x$:Yb^{2+} system ($x = 0.03$ and $x = 1$) with variation of the excitation energy. The data were obtained by simultaneously measuring the VL, EL, and their ratio VL/EL (while scanning the excitation wavelength). The absorption ranges of the various defect types with different numbers n of CN^- constituents are indicated by *gray bars*

wards lower excitation and EL energies, as discussed before. For comparison, we have included schematically the spectral ranges in which the various center types absorb. This makes it obvious that the step (indicated by arrow) can be attributed to a change from complexes with 2 CN^- ($n = 2$) to ones with $n > 2$. For energies below the step only centers with $n > 2$ absorb. Those centers obviously show a significantly increased E–V transfer efficiency. The highest efficiency is achieved for $n = 6$ in pure KCN. The lower efficiency at high excitation energies is caused by a "filter effect" of the inefficient complexes with $n = 0$, 1.

Dependence of Emission Lifetimes on the Number of CN^- Neighbors. The E–V transfer and the EL are competing processes and therefore the increase of the transfer efficiency should also be reflected in a reduced EL lifetime as shown in Fig. 9.24 in two ways. In Fig. 9.24a the luminescence intensity is shown for a constant CN^- concentration and different spectral ranges, and in Fig. 9.24b it is shown for a given spectral position (NIR) but an increasing CN^- concentration, and consequently higher average number n of CN^- ions within the complexes.

This is just a selection of the variations performed. All of the results show a common tendency:

- The E–V energy transfer efficiency η_{e-v} increases when more CN^- partners are included in the complex.
  While for 2 CN^- the relative efficiency is 30% with an absolute rate of 2×10^2 s^{-1}, the efficiency increases continuously with the number n of CN^- neighbors, reaching 100% efficiency and a rate of about 10^6 s^{-1} for KCN ($n = 6$).

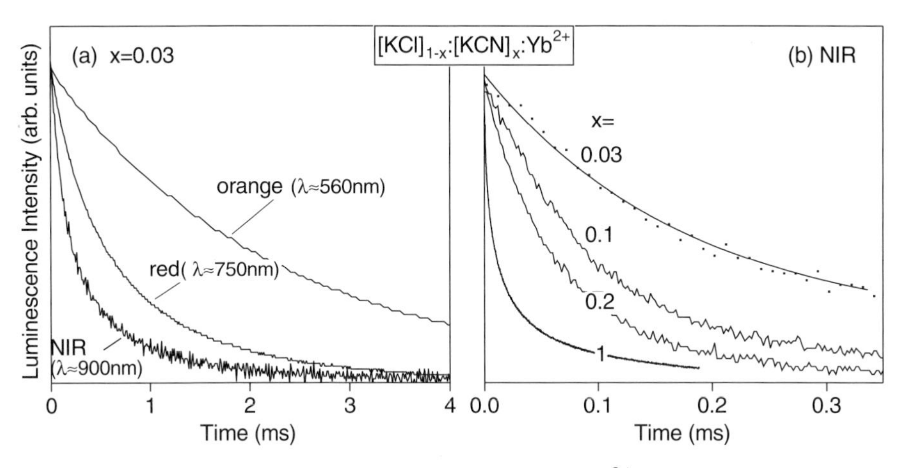

Fig. 9.24. EL decay curves in the $(KCl)_{1-x}$:$(KCN)_x$:Yb^{2+} system for (**a**) different spectral regions in a sample with $x = 0.03$ and (**b**) different center concentrations in the NIR spectral region

While these trends can be extracted fairly easily from experimental data, a selective study of complexes with a particular $n > 2$, is much harder due to their strong spectral overlap. One always deals with a mixture of centers with different n. For that reason the determination of the absolute E–V transfer rate as a function of n can only be obtained approximately from the multi-exponential decay curves. The obtained continuous increase of η_{e-v} with n is shown in Fig. 9.25. The most reliable values are the ones obtained for $n = 2$ and for $n = 6$ (KCN).

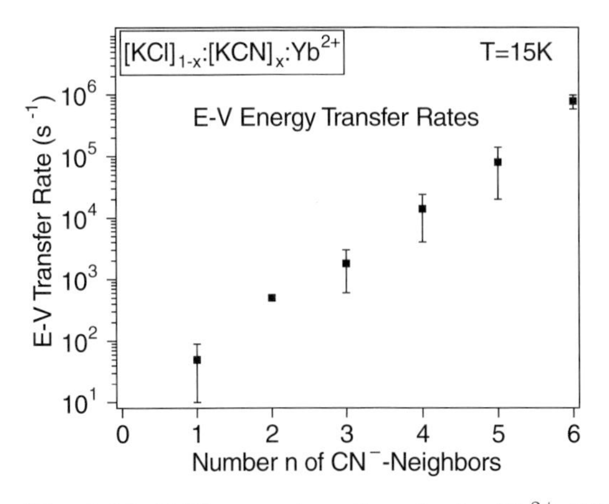

Fig. 9.25. E–V energy transfer rates for Yb^{2+}:$(CN^-)_n$ complexes in KCl as a function of the number n of CN^- neighbors

9.9 Properties of Yb^{2+} Ions with Excited CN^- Neighbors

The results presented so far for Yb^{2+}:$(CN^-)_n$ defect centers have shown that the electronic properties of the Yb^{2+} are changing drastically by the mere presence of CN^- molecules in the close neighborhood and by the relative orientation of the CN^- molecule. This sensitivity to the static environment raises the question of how the dynamics of the perturbation manifests themselves. More specifically, one might ask:

- How do the electronic properties change with excitation of the CN^- molecules?

In order to study this question, it would be most attractive to excite the CN^- molecule within a simple Yb^{2+}–CN^- defect pair, but such direct IR excitation is very hard to achieve experimentally. For that reason, another method was chosen, which involves optical excitation in the VIS and an energy transfer process to CN^-, and hence this method was limited to the more complex centers with more than one CN^- neighbor. The experimental pump–probe method, which is indicated schematically in the simplified energy-level scheme of Fig. 9.26, was essentially identical to the one used to determine the electronic–vibrational coupling of $F_H(CN^-)$ centers in Sect. 5.4. Indeed, the initial intention of this study was to determine the S_{CN} parameter for the case of Yb^{2+}-related centers as well, but it turned out that the electronic vibrational sidebands were obscured by another effect.

During laser excitation of Yb^{2+} ions in an ionic solid, several possible transient absorption signals may occur. These possibilities are listed below along with their characteristic properties, which helped us to identify the one relevant to us:

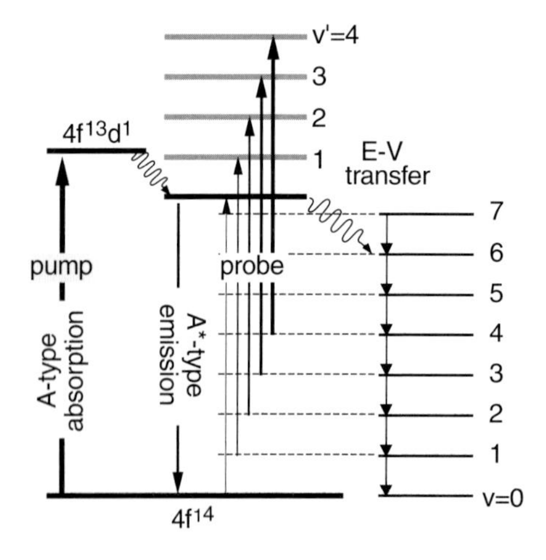

Fig. 9.26. Schematic model for explanation of the observed transient absorption

(a) Absorption from electronically excited states, which should disappear at the same rate as the electronic luminescence.
(b) Ionization of the Yb^{2+} ion and formation of Yb^{3+}, with fairly sharp and characteristic 4f $\leftrightarrow$ 4f transitions around 1.3 eV, as found in CaF_2 for instance [38].
(c) Ionization of the Yb^{2+} ion and formation of an electron-excess center involving another defect. The formation of such a new defect would be isotropic, and no polarization effect would be expected.
(d) Ionization of the Yb^{2+} ion and formation of an impurity-trapped exciton at the same site. Such a process has been observed in the anomalous emission of SrF_2 [20], and the resulting transitions depend strongly on the exciton bonding energies and thereby on the host material.
(e) Electronic–vibrational sidebands. The corresponding absorption peaks should show a regular structure (as in Fig. 5.6) reflecting the distance ($\sim$ 0.25 eV) between the vibrational levels.
(f) *Change of transition probabilities due to vibrational excitation of the* CN^- *ion.*

By a method of exclusion it was found that only the last plays a major role. The arguments will be presented in the following.

9.9.1 Experimental Results and Their Interpretation

The measured transient absorption (TA) and regular absorption spectra are shown in Fig. 9.27. Quite differently from the measurements on $F_H(CN^-)$ centers, the observed signals are very strong; the TA cross sections are of the same order as those of the regular absorption of the excited Yb^{2+}:$(CN^-)_{n\approx 2}$ centers. Moreover, in the spectral regions measured only positive TA signals were observed. This means that no significant bleaching of absorption takes

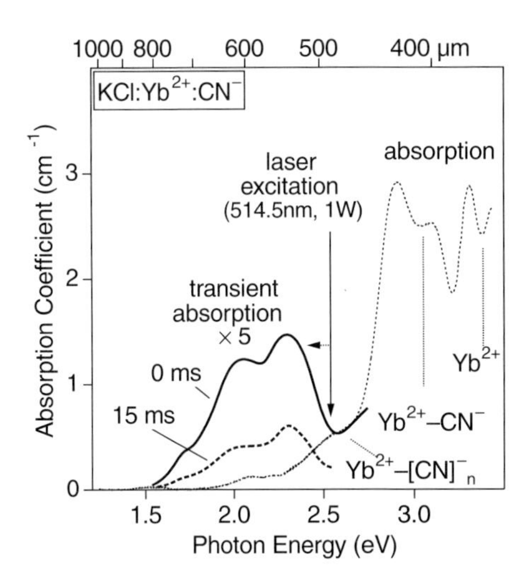

Fig. 9.27. Transient absorption spectra of Yb^{2+}:$(CN^-)_n$ centers in KCl just after (0 ms) and 15 ms after 488 nm laser excitation. For comparison, the regular absorption spectrum of the sample is shown

place and/or that this EA reduction is overcompensated by the appearance of a stronger absorption. As the number of Yb^{2+} ions does not change, this behavior can only be understood as indicating an increased EA cross-section. Although some structure can be found in the TA spectra, the spectral spacings are quite different from 0.25 eV and hence along with the complete absence of any reduction of the $\Delta v = 0$ transition strength, exclude the possibility that the electronic–vibrational sidebands are responsible for the TA signal.

The spectral position, a clear polarization dependence and an almost host-independent spectral shape exclude the possibilities (b)–(d) above. Furthermore, the rather slow reduction of the TA effect for KCl excludes option (a), leaving only (f) to be considered further. In order to test this option we described the dynamical behavior by means of a multilevel system, as depicted in Fig. 9.26. After electronic excitation into an A-type absorption band and fast relaxation into the lowest electronic state, either an A*-type emission or an E–V transfer takes place. After excitation, the molecule relaxes down the vibrational ladder, and does so partially radiatively giving rise to vibrational luminescence. The dynamics of this cycle have been extensively studied as described in Sect. 9.8, such that all relevant rates are known. To account for the TA effect, it was assumed that the electronic absorption of the Yb^{2+} ion is enhanced by a factor which depends on the vibrational excitation of the CN^- neighbors. Owing to the similarities of the TA spectra with the spectra for electronic luminescence, we considered the A*-type transitions especially. The level scheme in Fig. 9.26 is easily "translated" into a system of rate equations, and the time-dependent TA signal can be fitted by solving the equations numerically. Very good agreement was found by using just a v-dependent enhancement factor for the A*-type EA as a fitting parameter. To obtain this agreement, the plausible result that the higher excitations contribute more strongly to the effect had to be assumed. The observed subband structure in the TA spectra and the change in the relative strengths of the subbands with time are reflected in enhancement factors which depend on the spectral position. This behavior suggests that several different defect types participate in the TA effect.

Further confirmation of this model was obtained from the temperature dependence of the TA decay which follows very closely the corresponding dependence of the VL in KCl. An even better proof was obtained by variation of the host material. While for all hosts studied the electronic relaxation occurs on the same timescale ($\approx$ 3 ms), the vibrational relaxation is drastically different for different hosts. In NaCl, the nonradiative channels dominate and the VL lifetime is very short ($\sim$ 100 μs), while in KBr the VL has almost the full quantum efficency and has a lifetime of approximetely 20 ms. This strong variation in the VL decay rate can be clearly observed in the decay of the TA signal after the laser is switched off (see Fig. 9.28), giving a clear connection between the v-level population and the TA.

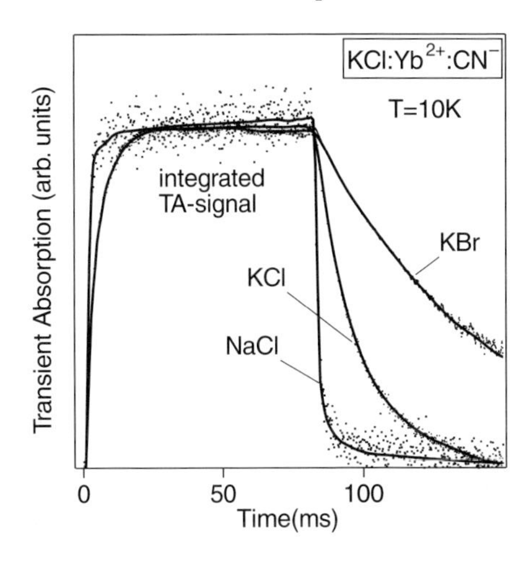

Fig. 9.28. Time dependence of the transient absorption spectra of Yb^{2+}:$(CN^-)_n$ centers in KCl, NaCl, and KBr under 488 nm laser excitation

9.9.2 Origin of Enhancement of Electronic Transition Probability

Despite the strong evidence that the observed enhancement is connected with a dynamic interaction of the Yb^{2+} ion with a vibrating molecular neighbor, one main question still needs to be addressed:

- What is the underlying mechanism of the observed enhancement effect?

In our attempt to answer this question, it is worth recalling that we have already found on two occasions that the strength of the A* absorption is very sensitive to the static and dynamic environment:

- The A*/A ratio of the integrated absorption cross section is strongly dependent on the host and defect type (see Sect. 2.2.1).
- The lifetime of the A*-type emission increases with temperature. The size of this effect increases with the number of CN^- neighbors.

It is apparent, especially from the latter observation, that dynamic interaction is able to increase the transition probability. A possible cause is a dynamic reduction of the local symmetry leading to a mixing of electronic states and thereby making the forbidden A*-type transitions allowed. This could be caused by the localized modes due to the CN^- neighbor, in addition to the phonon modes of the host lattice. While only low-frequency librational or translational modes of the CN^- can be thermally activated when the temperature is increasing, the stretch mode may also play a role in the TA measurement. However, owing to the partially nonradiative vibrational decay, the other modes are excited during the optical cycle as well. The interaction between the Yb^{2+} and CN^- ions could be based on both the electric and the elastic dipole of the molecule. The observation in $SrCl_2$ that the Yb^{2+} transitions are considerably influenced in position and strength by an applied stress field but very little by an electric field [39] makes the elastic interaction the more likely candidate. Moreover, it is known for the CN^- ion that its permanent electric dipole is very small, such that, for instance, in pure KCN the orientational ordering of the CN^- ions is based on the elastic interaction (see e.g. [40]).The relatively small TA effect for KBr, in which the nonradiative channel is very weak, and the appreciable effect in NaCl, in which the vibrational excitation is completely distributed into local and phonon modes, suggest that of the modes of the CN^- defects the low-frequency librational and translational modes play the dominant role in the enhancement. Within this interpretation, the origin of the increased emission probability at elevated temperatures (see Figs. 9.30 and 9.23) and the enhanced VA under laser excitation have the same origin and differ only in the method of excitation of the interacting modes which may occur either thermally, or optically via E–V transfer and vibrational relaxation.

For the determination of the electronic–vibrational coupling parameter S_{CN}, the TA measurements were a complete failure because no sidebands could be observed, suggesting that S_{CN} is fairly small, even smaller than for the F center cases. However, the finding that the presence of the CN^- ion enables an effective mixing of electronic states shows that the "promoting-mode factor" might be quite appreciable and compensate for the small S_{CN}. This and the consequences of all the other experimental results for the modeling of the E–V transfer will be discussed in the next section.

9.10 Putting It All Together: Comparing E–V Transfer Rates with Theoretical Models

A summary of the experimental results and their consequences answer most of the questions raised above as follows:

- Center complexes with $n = 2$ or more CN^- molecules exhibit E–V energy transfer. The transfer is more efficient the more the electronic emission is shifted to lower energy (i.e. the more CN^- molecules there are in the neighborhood of the Yb^{2+} ion). For $n = 2$ the absolute transfer rate is $W_{\mathrm{e-v}}^{\mathrm{tot}} \approx 100\ \mathrm{s}^{-1}$.
- In the case of low CN^- codoping (i.e. $n = 2$) the E–V transfer occurs predominantly into the levels $v = 4$, 5, 6, and 7.
- The electronic emission and the E–V transfer are competing processes from a common relaxed excited state.
- The E–V transfer efficiency stays almost constant when the temperature is varied, although the radiative transition probability of Yb^{2+} increases by several orders of magnitude. This is only possible if the E–V transfer rate is proportional to the radiative rate.
- The Huang–Rhys factor S_{ph} for the coupling of the 5d electron to the lattice is in the order of 5.

- No vibrational sidebands could be observed limiting the molecular Huang–Rhys factor S_{CN} to less than 0.01.
- A strong dynamic coupling of CN^--related modes to the electronic Yb^{2+} states is found, which enhances otherwise forbidden transitions.

On the basis of these experimental findings the applicability of the E–V transfer models described in Sect. 4 will be discussed in the following.

9.10.1 The Förster–Dexter Model and the Relative Transfer Rates

Concentrating first on the qualitative aspects, we apply the FD energy transfer model. Using the relation presented in Sect. 4.3.1, the measured emission spectra and the relative vibrational transition probabilities obtained from Fig. 3.1 with application of (9.16), we are able to determine the relative E–V transfer rates. In order to visualize the results, Fig. 9.29 shows the energy of the electronic luminescence and the vibrational excitation energies of the v-levels ($v = 1 \ldots 6$) on the same (vertical) energy scale. The dark bars on the vibrational levels indicate the distribution of the E–V transfer into the different v-levels, as determined from the evaluation of the time dependent measurements of spectrally resolved VL (Fig. 9.20). It is obvious that E–V transfer only takes place when there is at least some overlap between the

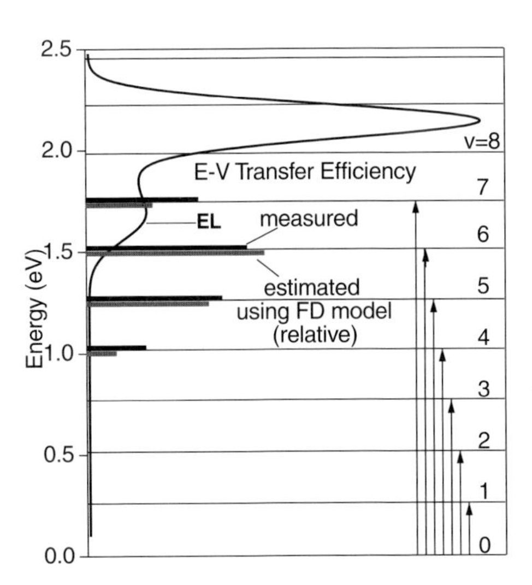

Fig. 9.29. Schematic illustration of the Förster–Dexter spectral-overlap model for Yb^{2+}:$(CN^-)_n$ in KCl. The EL and the CN^- vibrational levels, with some possible absorption transitions, are shown on a vertical energy scale. The measured relative v-dependent E–V efficiencies (obtained using fits from Fig. 9.20) are included as *dark bars* and the efficiencies estimated with the FD model are represented by *gray bars*

electronic emission and the vibrational states. However, the maximum of the E–V energy transfer distribution is shifted (relative to the red EL) towards lower energy. This is not surprising, because both the reduced energy denominator in (4.3) and the increased vibrational transition probability favor the lower vibrational levels, which can compensate for the smaller spectral overlap. The calculated relative E–V transfer rates (indicated as gray bars on the vibrational levels) agree quite well with the observations. In particular, the levels, $v = 5$ and 6 are predicted correctly as the main destination of the transfer. Besides this agreement at low temperature, the measured T-dependence of the transfer rate can easily be accounted for within the FD model because it predicts, according to (4.3) a rate $W_{\mathrm{e-v}}$ proportional to the radiative rate, just as we observed experimentally. A temperature-dependent spectral overlap term can account for the absence of a "perfect" proportionality of $k_{\mathrm{e-v}}$ and $k_{\mathrm{e}}^{\mathrm{rad}}$. Moreover, the observed tendency that the E–V transfer efficiency increases as the EL shifts to lower energy is, within the FD model, a direct consequence of the fact that the vibrational transition probability and therefore the integrated absorption cross section Q_{CN} increases towards lower vibrational levels and transition energies.

9.10.2 FD Model: Absolute Transfer Rates

Overall, the FD model accounts well for the *qualitative* aspects of our results but fails to explain the *quantitative* features. The absolute E–V transfer rate calculated using (4.3) comes out several orders (≈ 4) of magnitude too low. In the treatment of the FD model, we assumed that the defect partners interact with each other *only* via dipole–dipole interaction. Several experimental observations, however, show that this cannot be the case and that this assumption has to be removed. For instance, the presence of an electronic-defect neighbor can change Q_{CN}. This is also the case for the Yb^{2+}–CN^- defects as can be seen from the change of vibrational absorption strength observed in measurement of optically induced bistability as described in Sect. 9.4. An even stronger effect is expected in the lowest excited state from which energy transfer occurs if we use the simple model of Sect. 3.2.2, because this state is strongly polarizable owing to the energetic vicinity of states with different parity. Any increase in Q_{CN} will improve the agreement for the E–V transfer rate. However it remains doubtful whether a factor as high as 10^4 is possible, because no VL originating from molecules interacting with electronically excited Yb^{2+} ions could be observed, even under the most intense excitation. For that reason, the enhancement factors are limited to less than 100.

9.10.3 Horizontal-Tunneling Model: Relative Transfer Rates

As pointed out in Sect. 4.3.1, under most conditions it is possible to account for the relative E–V transfer efficiency within the horizontal-tunneling or supermolecule model whenever an agreement is achieved using the FD model.

The fitting is just a question of the choice of S_{CN} for the electron–vibrational coupling. In cases of very small values of S_{CN} – just as we encounter here – the Franck–Condon factors may be dominated by the contributions from the anharmonicity and are directly related to the transition probabilities according to (4.18). In the horizontal tunneling model, the only temperature dependence of the rate arises through the phonon occupation and therefore this model cannot account for the observed proportionality of the transfer rate to the electronic transition probability if a constant promoting-mode factor is assumed. If we recall that this factor describes the admixture of initial and final electronic states "promoted" by a vibrational mode, we find a clear similarity to the interpretation that we used in Sect. 9.9.2 to explain the increase of the electronic transition probabilities due to excited, localized, CN^--related modes. The only real differences lie in the type of states considered in the mixing. The lowest excited state mixes with ground state in one case and with other excited states in the other case. It is plausible to assume that both of these mixing effects have a similar (not necessarily equal) dependence on the degree of thermal excitation of the mode responsible. In this way the T-dependence can be explained at least qualitatively by a T-dependent promoting-mode factor which is dominated by a localized CN^--related mode.

9.10.4 Horizontal-Tunneling Model: Absolute Transfer Rates

Again, the real test of an E–V transfer model is the correct prediction of the absolute transfer rate. The lack of a detailed knowledge of electronic-vibrational coupling parameters and promoting-mode factors, however, leaves this task wide open to speculation. Without any doubt, parameters can be found which yield high enough rates. Indeed, promoting-mode factors of the order of those assumed for F centers and a coupling constant $S_{CN} = 0.01$ are sufficient to give high enough rates. However, both of those open parameters may be overestimated. For instance, using just the Franck–Condon factor due to the CN^- anharmonicity (and no shift of the potentials) gives rates which are too low. On the other hand, the promoting-mode factor might be underestimated in view of the strong dynamic mixing effects of states observed. In short: *The question of the rather high absolute* E–V *transfer efficiencies cannot be answered completely.*

9.11 Possible Application as a Phosphor

The visible appearance of the light emitted by an alkali halide sample codoped with Yb^{2+} and CN^- when it is excited by a blue laser ($\sim$ 450 nm) suggests that this material might be suitable as a phosphor. In particular, excitation with the recently developed blue LEDs and diode lasers is very attractive. In order to quantify this first impression, the author studied the electronic luminescence not only at low temperature but also at higher temperatures (up to room temperature). As first noticed by Müller et al. [41] and shown in Fig. 9.30b, the luminescence intensity of the Yb^{2+}–CN^- pair centers increases with an increase of temperature. As shown in Fig. 9.30a this is accompanied by an increase of the radiative transition rates, which becomes more pronounced with increasing number of CN^- neighbors. The physical background to this phenomenon which helps to increase the quantum efficiency of the emission, has been discussed in Sect. 9.9.2: thermal activation of localized, CN^--related modes mixes states of different parities. The initial idea which started the field, works in this defect system and might prove very useful in applications. Just as at low temperature the emission spectra change when the excitation energy and CN^- concentration are varied allowing one to tune the emission to a desired position. For one sample, the first possibility is shown in a contour plot of a CEES experiment (Fig. 9.31). The dark line indicates how the emission peak position moves as the excitation energy is changed. For the sample chosen here, a variation between 570 nm and 620 nm can be obtained.

While the CEES representation of the data is very informative for spectroscopists, for application as a phosphor the "color" of the emission is more important. Although color is a very subjective sensation, a whole field of

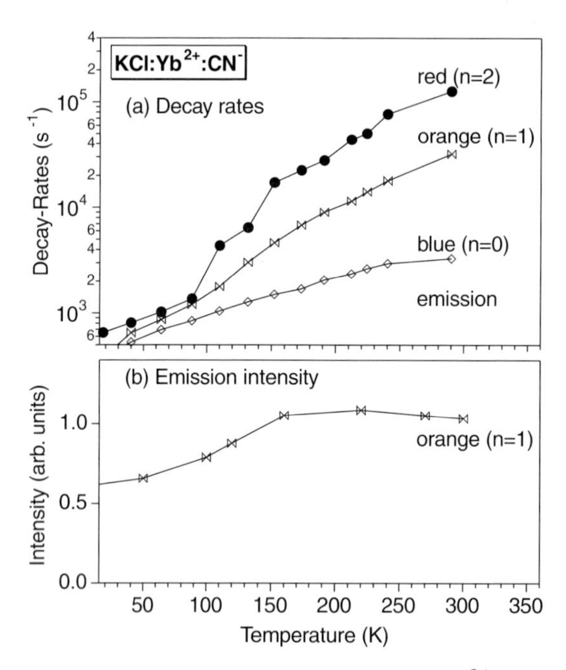

Fig. 9.30. Temperature dependence of Yb^{2+} luminescence in KCl: (**a**) decay rates of the blue, orange, and red luminescence, and (**b**) intensity of the orange luminescence

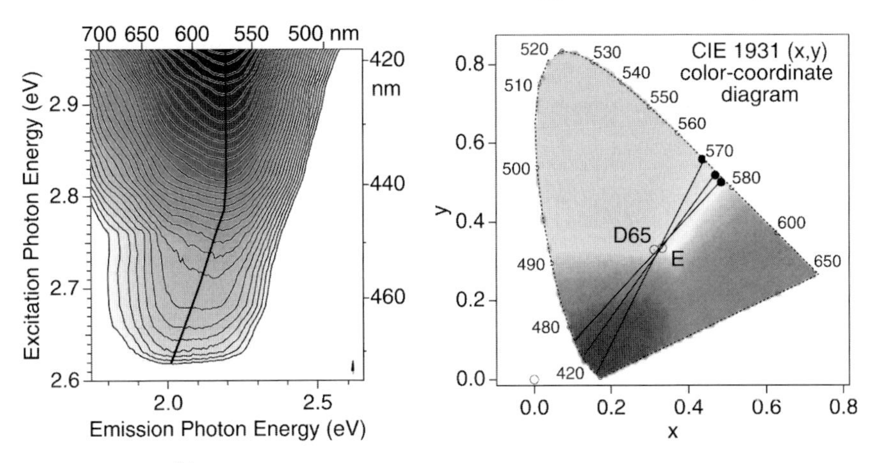

Fig. 9.31. Yb^{2+}:$(CN^-)_n$ centers in KCl at room temperature. *Left*: contour plot of CEES experiment. The *dark line* indicates the excitation-energy dependence of the peak emission energy. *Right*: color coordinate diagram. The • *symbols* show the x, y coordinates for different excitation energies; the *lines* indicate the possible color variation if the emission is mixed with the corresponding excitation light. The two ○ *symbols* indicate the D65 and E points, which correspond to a white color

science deals with the quantification of the impression of color for standard observers (see for example [42]). Several different methods exist. In the following, only the standard laid out by the CIE in 1931 ("CIE 1931") will be used. The standard observer in this method is described by three color-matching functions $\overline{x}(\lambda)$, $\overline{y}(\lambda)$, and $\overline{z}(\lambda)$ which are listed for example on the CIE home page.[4]

A measured spectrum $I(\lambda)$ can be converted into color coordinates x, y using the following equations:

$$
\begin{aligned}
X &= \sum_i \overline{x}(\lambda_i)\, I(\lambda_i)\,, \\
Y &= \sum_i \overline{y}(\lambda_i)\, I(\lambda_i)\,, \\
Z &= \sum_i \overline{z}(\lambda_i)\, I(\lambda_i)\,, \\
x &= \frac{X}{X+Y+Z}\,, \\
y &= \frac{Y}{X+Y+Z}\,, \\
z &= \frac{Z}{X+Y+Z}\,.
\end{aligned}
\tag{9.18}
$$

[4] http://www.cie.co.at/cie/home.html

The values of x, y calculate with (9.18) can be depicted in a color coordinate diagram. The color matching functions and the condition that $x + y + z = 1$ limit the coordinate space, and all visible colors are located within a certain section of this space as indicated in Fig. 9.31. The (x, y) values for our spectra are located on the edge of this section.

The possibility of mixing colors to produce new ones is important for application. This can be illustrated in a color coordinate diagram by a line which connects the points corresponding to the (x, y) coordinates of the two color sources. If we do that in our case and mix the emission and the excitation light, we are able to move along the solid straight lines in Fig. 9.31, which cover a wide range of colors, and in the middle pass through a range in which the color impression is "white". This region is indicated by two standard points (D65 and E), one being the color value for the D65 stimulant, which corresponds closely to a black body at 6500 °C and is very close to daylight in northern Europe and the other point E is the point of equal energy ($x = 0.3333$, $y = 0.3333$). On the basis of this consideration, it is clear that in principle it should be possible, by proper mixing of the light from a 450 nm LED and the emission from the Yb^{2+}–CN^--type defects to realize "white" LEDs similar to the ones already commercialized by Nichia and other companies [43]. As our systems share a high quantum efficiency with the organic dyes and Ce^{3+}-doped glasses used in the Nichia LEDs, the possibility of changing the emission wavelength not only through the excitation wavelength but also through the $[CN^-]$ concentration makes this system quite attractive. Besides the application as an incoherent emitter described above, laser operation might be possible as well, especially since some initial (unconfirmed) measurements of optical gain have been reported [44].

References

1. C. P. An, V. Dierolf, and F. Luty, Phys. Rev. B **61**, 6565 (2000).
2. J. Ortiz-Lopez and F. Luty, Phys. Rev. B **37**, 5452 (1988).
3. J. O. Rubio, J. Phys. Chem. Solids **52**, 101 (1991).
4. M. Schrempel, W. Gellermann, and F. Luty, Phys. Rev. B **45**, 9590 (1992).
5. C. P. An (unpublished).
6. P. Beneventi, R. Capelletti, L. Kovács, A. M. L. Manotti, and F. Ugozzoli, J. Phys.: Condens. Matter **6**, 6329 (1994).
7. P. Dumas, Y. J. Chabal, and G. Higashi, Phys. Rev. Lett **65**, 1124 (1990).
8. H. Bilz and W. Kress, *Phonon Dispersion Relations in Insulators*, Springer Series in Solid-State Science, Vol. 10 (Springer, Berlin, 1979).
9. G. Baldacchini, S. Botti, U. M. Grassano, L. Gomes, and F. Luty, Europhys. Lett. **9**, 735 (1989).
10. H. Söthe, J.-M. Spaeth, and F. Luty, J. Phys.: Condens. Matter **5**, 1957 (1993).
11. V. Dierolf and F. Luty, Phys. Rev. B **54**, 6952 (1996).
12. F. Luty, in *Physics of Color Centers*, edited by W. Fowler (Academic Press, New York, London, 1968) p. 181.

13. H. Schläfer and G. Gliemann, *Einführung in die Ligandenfeldtheorie* (Akademische Verlagsgesellschaft, Frankfurt am Main, 1967).
14. C. Jørgensen, *Modern Aspects of Ligand Field Theory* (North-Holland, Amsterdam, London, 1971).
15. L. Niinistö and M. Leskelä, in *Handbook on the Physics and Chemistry of Rare Earths*, edited by J. K. A. Gschneidner and L. Eyring (North Holland, Amsterdam, 1987), Vol. 9.
16. S. W. Bland and M. J. A. Smith, J. Phys. C: Solid State Phys. **18**, 1525 (1985).
17. T. S. Piper, J. P. Brown, and D. S. McClure, J. Chem. Phys. **46**, 1353 (1967).
18. M. V. Eremin, Opt. Spectrosc. **29**, 53 (1970).
19. T. Tsubui, H. Witzke, and D. S. McClure, J. Lumin. **24&25**, 305 (1981).
20. B. Moine, C. Pedrini, D. S. McClure, and H. Bill, J. Lumin. **40&41**, 299 (1988).
21. S. Lizzo, A. Meijerink, G. J. Dirksen, and G. Blasse, J. Lumin. **63**, 223 (1995).
22. S. Lizzo, A. Meijerink, and G. Blasse, J. Lumin. **59**, 185 (1994).
23. D. S. McClure and C. Pedrini, Phys. Rev. B **32**, 8465 (1985).
24. E. König and S. Kremer, *Ligand Field Energy Diagrams* (Plenum, New York, 1977).
25. E. K. Ivoilova and A. M. Leushin, Sov. Phys. Solid State **22**, 601 (1980).
26. B. Z. Malkin, in *Spectroscopy of Solids Containing Rare Earth Ions*, edited by A. A. Kaplyanskii and R. M. Macfarlane (Elsevier Science, Amsterdam, 1987), Vol. 21 (Chap. 2, p. 13).
27. I. Colquhoun, N. Greenwood, I. J. McColm, and G. E. Turner, J. Chem. Soc. Dalton Trans. **13**, 1337 (1972).
28. P. Souchko, A. Shluger, C. R. A. Catlow, and R. Baetzold, Radiat. Eff. Defects Solids **151**, 215 (1999).
29. A. Shluger, private communications (1998).
30. V. Dierolf and F. Luty, Rev. of Solid State Sci. **4**, 479 (1990).
31. K. Koch, Y. Yang, and F. Luty, Phys. Rev. B **29**, 5840 (1984).
32. J. T. McWhirter, U. Happek, and A. J. Sievers, J. Lumin. **60&61**, 846 (1994).
33. D. Samiec, Ph.D. thesis, Universität GH Paderborn, 1997.
34. E. Gustin, M. Leblans, A. Bouwen, and D. Schoemaker, Phys. Rev. B **54**, 6963 (1996).
35. T. Tsubui, D. S. McClure, and W. C. Wong, Phys. Rev. B **48**, 62 (1993).
36. F. Luty and V. Dierolf, in *Defects in Insulating Materials*, edited by O. Kanert and J.-M. Spaeth (World Scientific, Singapore, 1993), p. 17.
37. V. Dierolf and F. Luty, in *Defects in Insulating Materials*, edited by O. Kanert and J.-M. Spaeth (World Scientific, Singapore, 1993), p. 559.
38. B. Henderson and G. Imbusch, *Optical Spectroscopy of Inorganic Solids* (Oxford Science, Oxford, 1989).
39. A. Kaplyanskii and P. Smolyanskii, Opt. Spectrosc. **40**, 528 (1976).
40. F. Luty, in *Defects in Insulating Crystals*, edited by V. Turkevich and K. Shvarts (Springer, Berlin, 1981), pp. 69–89.
41. M. Müller, J. F. Fabris, A. C. Hernandes, and M. Siu-Li, J. Lumin. **59**, 289 (1994).
42. G. Wyszecki and W. Stiles, *Color Science*, 2nd ed., Wiley Series in pure and applied optics, Vol. XV (Wiley, New York, 1982).
43. S. Nakamura and G. Fasol, *The Blue Laser Diode: GaN Based Emitters and Lasers* (Springer, Berlin, 1997).
44. M. Müller, J. L. Fabris, A. C. Hernandes, and M. Siu-Li, unpublished (1994).

10 Europium and CN^- Molecules

Among the divalent rare-earth defects, the isolated Eu^{2+} ion has been studied most intensively in the past in numerous ionic crystals, including the alkali halides [1]. Owing to the high quantum efficiency of its blue emission, it has been considered as an active ion for laser operation and other applications based on its luminescence properties (e.g. X-ray storage, optical memory [2], and dosimeters [3, 4]). All of these applications use the blue Eu^{2+} emission for probing and would profit from a further increase of the quantum efficiency and a shift of the emission energy to longer wavelengths, into the spectral range ($\approx$ 700 nm) in which simple Si detectors and CCD arrays are most sensitive. In this chapter, it will be shown that this can indeed be made possible by doping with CN^-.

In our studies of Eu^{2+}:$(CN^-)_n$ complexes, many similarities to the case of Yb^{2+}-related defects appeared, so we shall present fewer details here.

10.1 Eu^{2+}:$(CN^-)_n$ Complexes

Adding CN^- as a co-dopant to the melt in the crystal growth process changes the spectral positions of the Eu^{2+} transitions, as shown in Fig. 10.1 for KCl. This behavior is reflected in a color change of the melt and the crystal: while samples doped only Eu^{2+} are transparent, the codoped samples are yellowish. Comparing these changes in the experimental findings for Eu^{2+} with those for Yb^{2+}, the following similarities and minor differences were found:

- With increasing CN^- concentration the absorption bands shift towards lower energies; however, this shift is less pronounced than for Yb^{2+}, making a decomposition difficult.
- Studies with different concentrations of Eu^{2+} and CN^- suggest that the formation of Eu^{2+}:$(CN^-)_n$ complexes is not very efficient, such that the complex concentrations are lower than expected from statistics. Formation of Eu^{2+} aggregates seems to be more favorable (see e.g. [1, 5]).
- Excitation in the new absorption bands produces strong emission bands which are shifted relative to those for the isolated ion by similar amounts to the shifts for Yb^{2+}.

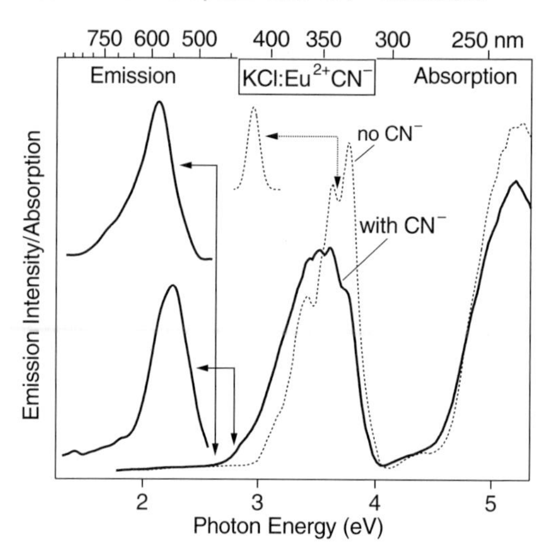

Fig. 10.1. Absorption and emission spectra for Eu^{2+}:$(CN^-)_n$ defects in KCl. The positions of the excitation for the two emission spectra are indicated by *arrows*. The spectra without CN^- are shown by *dashed lines* for comparison

- The emission is very inhomogeneous, as can be seen from the combined excitation–emission measurement shown in Fig. 10.2. The contour plot is almost indistinguishable from that measured for Yb^{2+}-doped samples (Fig. 9.3), reflecting the strong similarities in the emission behavior.

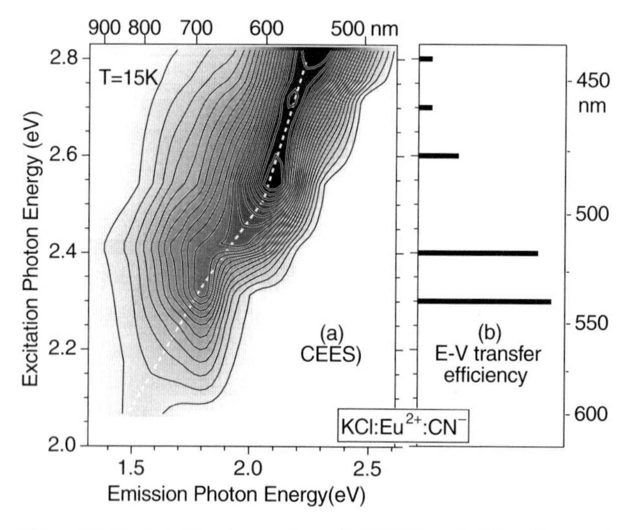

Fig. 10.2. (**a**) Contour plot of CEES and (**b**) relative E–V transfer efficiency for several excitation photon energies for Eu^{2+}:$(CN^-)_n$ defect complexes in KCl at $T = 15$ K

- The 4f ↔ 4f transitions when studied by two-photon absorption are hardly changed [6], indicating that the interaction effects are small for the 4f electron configuration. We shall come back to this issue in the next chapter, in which we compare interaction effects for the 5d and 4f states of Sm^{2+} ions.

An E–V transfer from excited Eu^{2+} 5d levels into the stretch-mode of the CN^- ion in KCl, KBr, and KI was first found by Naber [7] in measurements of the vibrational luminescence and has been studied further by the author [8]. The VL spectrum for KBr is shown in Fig. 10.3 in which, in addition to VL responses from isolated CN^- molecules excited via V–V transfer, VL from at least two types of Eu^{2+}-perturbed CN^- molecules can be identified. For the dominant response (type I), the transition energy is higher than that for to the isolated molecule. The results in the other host materials are similar and the spectral positions of the VA lines detected are listed in Table 10.1.

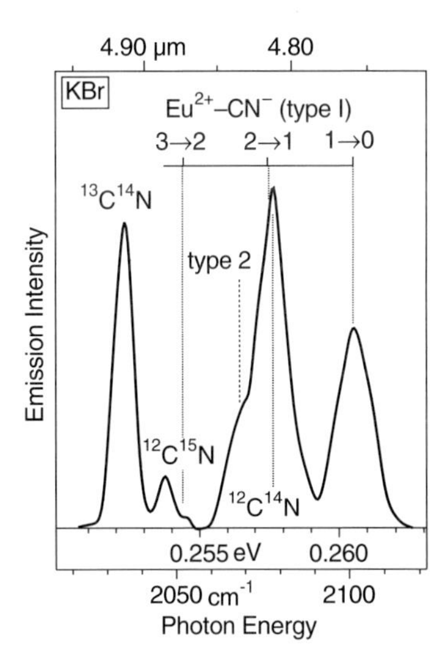

Fig. 10.3. Vibrational luminescence spectra of Eu^{2+}:$(CN^-)_n$ defects in KBr at $T = 4$ K [7]

Table 10.1. Vibrational frequencies (in cm^{-1}) of Eu^{2+} related CN^- defects in K halides obtained from vibrational luminescence [7]

Host			KCl	KBr	KI
Isolated CN^-		$\overline{\omega}_{\text{isol}}$	2088	2078	2067
Eu^{2+}-related	Type I	$\overline{\omega}_{\text{Eu}_1}$	2100	2101	2088
	Type II	$\overline{\omega}_{\text{Eu}_2}$	–	2070	2059
$\Delta_{\text{type1}} = \overline{\omega}_{\text{Eu}_1} - \overline{\omega}_{\text{isol}}$			12	23	21
$\Delta_{\text{type2}} = \overline{\omega}_{\text{Eu}_2} - \overline{\omega}_{\text{isol}}$			–	−10	−8

The VL signals are about one order of magnitude weaker than for Yb^{2+}, which most likely is related to the smaller number of complex centers formed in thermal equilibrium. Measurements of the integrated VL strength reveal (as indicated in Fig. 10.2) an increase of the E–V transfer efficiency towards lower excitation and luminescence energies.

10.2 Possible Application as a Phosphor

Similarly to the results for the Yb^{2+} doping, a very strong emission persists up to room temperature, making the system attractive for application as a phosphor. The absorption band is well suited to excitation at around 450 nm, a spectral range for which "blue" LEDs are commercially available. The corresponding emission, shown in Fig. 10.4 (left), has an intense green to yellowish color and can be mixed to the blue excitation and produce white light very close in its color coordinates to the D65 reference point. From the visible impression, this light is even brighter than for Yb^{2+}-doped samples, suggesting that the quantum efficiency reaches almost unity. Such a high value is not uncommon for Eu^{2+}-doped compounds.

10.3 Summary and Interpretation of Experimental Results

Despite the fact that much less detailed information (compared with the Yb^{2+} case) has been obtained about the present defect system, the basic results are very clear:

- The Eu^{2+} electronic transitions are shifted to lower energies by the interaction with CN^- neighbors.
- The shifted electronic emission has a very high quantum efficiency and can be excited effectively using a blue LED (450 nm).
- The CN^- vibrational transitions are shifted to higher energies by the interaction with a Eu^{2+} neighbor.
- E–V energy transfer occurs with increasing efficiency as the excitation wavelength is increased, but is weaker overall than for Yb^{2+} in samples doped with similar amounts of CN^- and rare-earth ions.

These results strongly resemble those found for the Yb^{2+}:$(CN^-)_n$ complexes and therefore their interpretation is analogous, an obvious consequence of the similarities between a half-filled (Eu^{2+}) and a completely filled (Yb^{2+}) 4f shell. Without going through the same arguments again, we shall again assume a charge redistribution away from the CN^- ion to the 5d orbitals of the Eu^{2+} ion because both the redshift and the blueshift of the electronic and vibrational transitions, respectively, can be explained on this basis. This strong

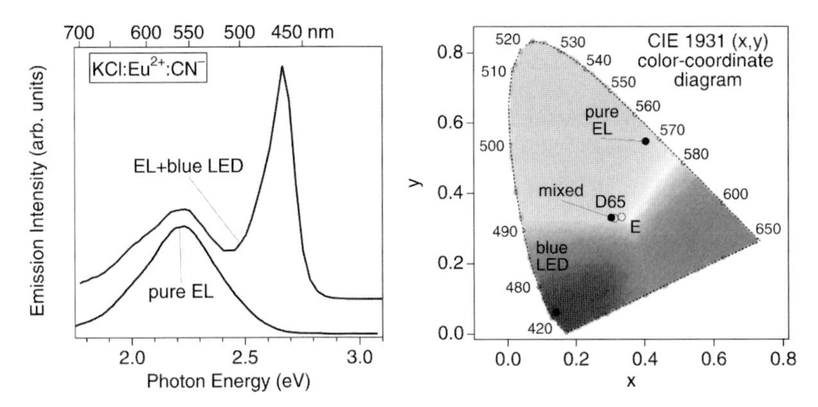

Fig. 10.4. KCl:Eu^{2+}+2% KCN at room temperature. *Left*: EL spectra under excitation with a blue LED. In one of the spectra, not all of the excitation light was absorbed. The "pure EL" was taken in a 90° geometry. *Right*: color coordinate diagram for the emission spectra shown and the spectrum of the blue LED (•). The region of white light is marked by the D65 and E reference points (○)

interaction should again be reflected (although it has not been measured) in an efficient dynamic mixing of electronic states and in a promoting-mode factor that allows a rather efficient E–V transfer despite the otherwise unfavorable conditions of rather small electron-phonon coupling and a large energy gap. However, the charge redistribution and the effect of bonding between the defect partners, which should be present just as in the case of Yb, make the treatment in terms of two individual interacting defects only a crude approximation and its conclusions somewhat questionable.

References

1. J. O. Rubio, J. Phys. Chem. Solids **52**, 101 (1991).
2. H. Nanto, K. Murayama, T. Usuda, F. Endo, Y. Hirai, S. Taniguchi, and N. Takeuchi, J. Appl. Phys. **74**, 1445 (1993).
3. R. Melendrez, R. Perez-Salas, L. P. Pashchenko, R. Aceves, T. M. Piters, and M. Barboza-Flores, Appl. Phys. Lett. **68**, 3398 (1996).
4. I. A. de Carcer, F. Cussó, F. Jaque, E. Espana, T. Calderon, G. Lifante, and P. D. Townsend, J. Phys. D: Appl. Phys. **26**, 154 (1993).
5. F. Mugenski and R. Cywinski, Phys. Status. Solidi B **125**, 381 (1984).
6. F. M. M. Yasuoka, J. C. Castro, and L. A. O. Nunes, Phys. Rev. B **43**, 9295 (1991).
7. A. Naber, Ph.D. thesis, Westfälische Wilhelms-Universität Münster, 1993.
8. V. Dierolf (unpublished).

11 Samarium and CN^- Molecules

The electronic–molecular defect systems considered so far (F center–CN^-, F center–OH^-, Yb^{2+}–$(CN^-)_n$) exhibit pronounced mutual coupling effects. Common to all these systems is a relatively strong electron–phonon coupling of the relevant electronic defect states, which is reflected in the corresponding Huang–Rhys factor S (e.g. $S = 10$–100 for F centers). This apparent connection between the coupling to the vibrational modes of the lattice and the coupling to those of the neighboring molecular defects has not been treated so far. To do this, we still need to consider defect systems for which the relevant electronic states are only weakly influenced by the surrounding ions. As an representative of that group, we discuss the Sm^{2+} ion and its interaction with CN^- molecules.

11.1 Introduction

Some possible and very interesting candidates for investigation of the interaction between molecules and ions with a weak electron–phonon coupling would be the trivalent rare-earth ions (e.g. Er^{3+} and Yb^{3+}), with their transitions between well-shielded 4f states, which have been used as laser-active ions for various applications. Unfortunately, it is in general not possible to introduce those ions into alkali halides, in which all the studies described above have been carried out. As an alternative, divalent rare-earth ions, such as Eu^{2+} and Sm^{2+}, which can be incorporated quite easily and have been well characterized already as individual defects [1], can be used. While the $4f \rightarrow 4f$ transitions of Eu^{2+} ions with CN^- neighbors have been studied by two-photon absorption [2], revealing essentially no interaction effect, we chose here the Sm^{2+} ion. This ion is of particular interest because some excited levels from the 5d and the 4f configurations are energetically very close, so that the interaction effects for weakly (4f) and fairly strongly (5d) electron–phonon-coupled states can be compared. We shall focus on the following questions:

(a) Regarding the spectral position and electron–phonon coupling, how are the different types of Sm^{2+} states influenced by the presence of CN^- neighbors?

(b) Can E–V energy transfer originating from the well shielded 4f states take place and how is the process altered when the strongly electron–phonon-coupled 5d states are involved?

In general, much less detail can be obtained, since all observed interaction effects are much smaller, as intended. Only through the application of very sensitive techniques such as CEES could important results be found.

11.2 Complexes Involving a Single CN^- Molecule

11.2.1 Spectroscopic Characterization

Guided by the results for the Yb^{2+}-related complexes, we start our consideration of the Sm^{2+}:$(CN^-)_n$ complexes with samples which have been codoped with fairly low amounts ($< 1\%$) of CN^- in order to obtain an understanding of the simplest complexes which involve only a single CN^- molecule. In Fig. 11.1, emission spectra for a KCl sample doped with both 0.1% Sm^{2+} and 0.5% CN^- are shown. For excitation at around 637 nm in the region of the zero-phonon line of the $^7F_0 \to 5d$ transition, the emission spectrum looks at

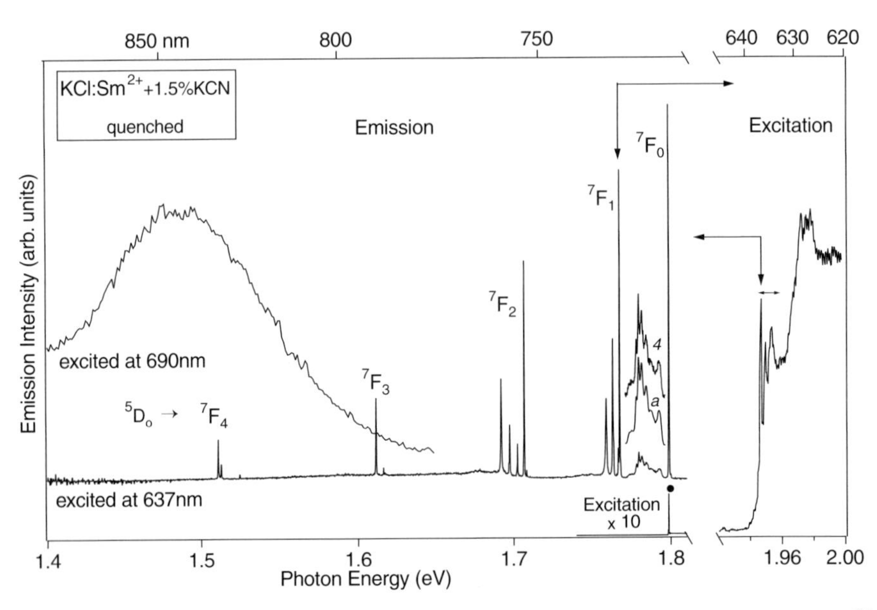

Fig. 11.1. Emission (excited at 690 nm and 637 nm) and excitation spectra of Sm^{2+} in KCl:Sm^{2+}+1.5% KCN. The sample was quenched from 600 °C just before the measurement. The positions of probing and exciting of the emission are indicated by *vertical arrows*. The vibrational sidebands of the $^5D_0 \leftrightarrow {}^7F_0$ transition are shown expanded for excitation of sites *a* and 4 at 1.7984 eV and 1.7980 eV, respectively. The *horizontal arrow* and the point • indicate the spectral ranges used in Fig. 11.2

first sight like that found for samples without CN^-. This spectrum consists of several groups of sharp lines due to emission from the $^5D_0$ state to the crystal-field-split ^7F_j states. Similarly, almost no change is seen in the excitation spectrum of the emission (probed at the spectral position indicated) for excitation in the $^7F_0 \rightarrow ^5D_0$ and the $^7F_0 \rightarrow 5d$ transitions. Obviously, the isolated Sm^{2+} ion is still dominant in the samples containing small amounts of CN^-, and a more detailed spectroscopic study is required to find the CN^--related centers.

A closer look at the emission spectra reveals a substructure within the $^5D_0 \rightarrow ^7F_j$ lines, which is present only for the codoped samples. The samples were studied further by CEES in two different spectral regions of the excitation energy (Fig. 11.2):

- interconfigurational (4f→5d) and
- intraconfigurational (4f→4f).

In order to allow reliable and easy comparison, special care was taken to perform the thermal quenching before measurements identically for the samples with and without CN^- doping. In the middle of Fig. 11.2, the same spectral ranges as in Fig. 2.7 are depicted. By comparison, we can find several extra sites (1–5) which appear only in CN^--doped samples, but do so even at low CN^- concentration. The peaks increase linearly relative to that of the main isolated Sm^{2+} site *a* with increasing CN^- doping. Hence, these peaks are unambiguously related to center complexes involving a CN^- molecular ion. Owing to their spectral closeness to the main center *a*, the assignment of peaks 6 and 7 is less certain. After identification of the various Sm^{2+}–CN^- centers in one spectral range, the regions of the other $^5D_0 \leftrightarrow ^7F_j$ emission transitions could be investigated by CEES to find the site-specific energies of the ^7F_j states, and we list those energies in Table 11.1. The differences in transition energies found are all of the order of 1 meV or less.

Despite its incompleteness (due to the small spectral overlap of the transitions), Table 11.1 represents the "fingerprints" of the various center types, which can be used to investigate them further *individually*.

For instance, we can study the electron–phonon coupling of the 4f states, which becomes apparent as vibrational sidebands, most pronounced next to the (zero-phonon) $^5D_0 \rightarrow ^7F_0$ emission transitions. For that purpose, the excitation was performed "resonantly" into this transition, i.e. at 1.7984 eV for the isolated Sm^{2+} defect *a* and at 1.7980 eV for the CN^--perturbed Sm^{2+}defect (predominantly in site 4). The resulting slightly Stokes-shifted phonon-assisted emission bands are shown in Fig. 11.1 on a vertically extended scale and labeled *a* and 4, respectively. We find that the shape and spectral position do not depend on the type of center excited. No indication of a coupling to a localized mode (as found for $F_H(CN^-)$ can be seen. Similarly, no resonance-enhanced Raman signal corresponding to the CN^- vibrational mode could be found. These two observations give clear evidence that the coupling of the 4f states to their close surroundings (in our case the CN^-

ion) is very weak, just as already found for the isolated Sm^{2+} defect [3, 4]. Furthermore, there is no evidence of E–V energy transfer.

Turning our attention from the 4f states to the 5d states, we can use the fingerprints to interpret the CEES results for excitation energies in the range of the vibrational substructure belonging to the interconfigurational 4f→5d transition. In the corresponding contour plot, shown in Fig. 11.2 (top), characteristic peaks appear for certain combinations of excitation and emission energies, which can be assigned to the different Sm^{2+} center types by comparison with the plot shown in the middle of Fig. 11.2. By looking for the common emission energies, the peaks originating from centers of type *a*, *d*, and 3–7 can easily be assigned and the energy of the corresponding zero-

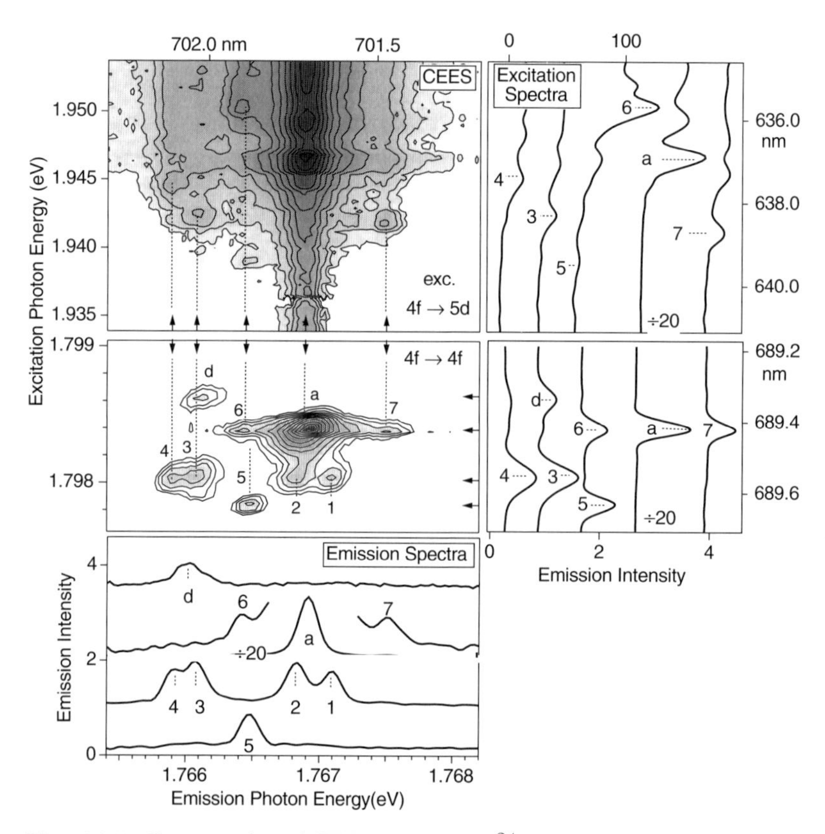

Fig. 11.2. Contour plot of CEES for KCl:Sm^{2+}+1.5% KCN under excitation (*top*) in the region of the vibrational phonon sidebands of the interconfigurational 4f→5d transitions, and (*middle*) in the region of the intraconfigurational $^7F_0 \rightarrow ^5D_0$ transition. The CN^--related sites are labeled with numbers 1–7. The sites *a*, *d* already present in the absence of CN^- are labeled according to [5]. The emission and excitation spectra for the spectral positions indicated by *arrows* are shown at the *bottom* and the *right*

phonon $^5D_0 \rightarrow 5d$ transition can be determined; this has been included in Table 11.1. The variations in the transition energies are about 10 meV. For sites 1 and 2 an identification is not possible since the emission energy is hardly distinguishable from that of the strongly emitting site *a*.

It should be noted at this point that all information provided by the CEES measurements is also contained in the individual (conventional) excitation and emission spectra, a few of which are shown in Fig. 11.2 for excitation and emission at the spectral positions indicated by the arrows. Although these positions have been chosen to be optimal for a particular defect site, the spectra look very complicated, but the energy positions can still be found (as indicated).

11.2.2 Preliminary Center Model

Although the spectroscopic results presented above are not suitable by themselves to identify the microscopic configurations of these centers, the number of different sites and the CN^- concentration dependence suggest that these are simple Sm^{2+}–CN^- complexes with different arrangements and orientations of the CN^- ion relative to the Sm^{2+} ion and the required charge-compensating cation vacancy, similar to what we have already seen in the case of Yb^{2+}–CN^- complexes (see Fig. 9.11).

Table 11.1. Spectral positions in eV of various Sm^{2+} and Sm^{2+}–CN^- transitions for different sites. In the cases where no value is given, no unambiguous assignment was possible

	Site a	Site d	Site 1	Site 2	Site 3	Site 4	Site 5	Site 6	Site 7
$^7F_0 \rightarrow 5d$	1.946	1.952	–	–	1.942	1.945	1.939	1.950	1.942
$^5D_0 \rightarrow {}^7F_0$	1.7984	1.7986	1.7987	1.7980	1.7980	1.7980	1.7978	1.7984	1.7984
$^5D_0 \rightarrow {}^7F_1$	1.7669	1.7662	1.7670	1.7668	1.7660	1.7658	1.7665	1.7664	1.7675
	1.7626	1.7650	–	1.7626	–	1.7622	1.7622	–	–
	1.7587	1.7577	–	1.7586	–	1.7579	1.7582	–	–
$^5D_0 \rightarrow {}^7F_2$	1.7063	1.7067		–	1.7066	–	1.7061	1.7059	–
	1.7020	1.7016	–	1.7021	–	1.7018	1.7015	–	–
	1.6969	1.6980	–	1.6967	–	1.6964	1.6965	–	–
	1.6915	1.6929	–	1.6915	–	1.6908	1.6909	–	–
$^5D_0 \rightarrow {}^7F_3$	1.6113	–	–	1.611	–	1.611	1.6108	–	–
	1.6111	–	–	1.6108	–	1.6108	1.6106	–	–
$^5D_0 \rightarrow {}^7F_4$	1.5132	–	–	1.5131	–	1.5128	1.5126	–	–
	1.5113	–	–	–	–	–	–		
	1.5110	–	–	1.5109	1.5107	–	1.5102	–	–

11.2.3 Energy-Level Scheme

As we have seen, the emission behavior of these centers is essentially the same as for the isolated defect so that the energy-level scheme of Fig. 2.5 can be used. This means that we are dealing with a 4f-type state as the lowest relaxed excited level, which should be the relevant level for the possible E–V energy transfer discussed later on. The observed spectral shifts in both the 4f and the 5d states are very small; they are even less pronounced than those seen for some simple Sm^{2+} configurations [5] and can easily be accounted for by small differences in the crystal field which the Sm^{2+} ion experiences. This change could be caused by the weak permanent electric dipole moment of the CN^- ion and amplified by a displacement dipole due to an off-center CN^- position. The electric fields from dipoles are smaller than or similar magnitude to those induced by charge-compensating vacancies and other Sm^{2+}–vacancy dipoles. Further interaction mechanisms such as covalency effects and electron exchange between the Sm^{2+} and CN^- ions, which are important for the much more pronounced shifts in Yb^{2+}–(CN^-) centers, apparently do not play a major role. This quite regular behavior is further reflected in the ratio (a factor of 10) of spectral shifts observed for the 4f and 5d states, which is commonly found in other rare-earth defect systems as well [6].

11.3 Complexes Involving Several CN^- Molecules

Similarly to the situation for Yb^{2+}:$(CN^-)_n$ complexes, more defect types appear when the CN^- concentration is increased. Their behavior is much more complicated and will be summarized in the following.

Even in moderately CN^--doped samples, we find, besides the sharp emission peaks described above, a much broader emission response, which can be observed for the whole excitation range that we studied (620–700 nm). An example is the spectrum obtained with 690 nm excitation shown in Fig. 11.1, a spectral position which does not coincide with any of the sharp spectral features of the spectra discussed above. The emission band, which has some faint sharp features on top, is strongly dependent in strength and spectral shape on the excitation position, suggesting a strong inhomogeneous broadening. It shifts to lower energy as the excitation wavelength increases. When the CN^- concentration is varied, it is found that this band is totally absent for samples not doped with CN^-. When the CN^- concentration is increased, the emission increases superlinear in its integrated intensity and shifts, for constant excitation energy to lower energies. These observations suggest, in analogy to the observation for other rare-earth defect systems, that complexes involving more than one CN^- ion are responsible for this type of emission.

11.3.1 Energy-Level Scheme

The spectral position and general appearance of the emission response resemble (apart from the less pronounced substructure) the spectra observed for Sm^{2+} complexes, so that we suspect that the lowest 5d level is shifted below the $^5D_0$ state. The amount of this shift depends on the number of CN^- neighbors included in the Sm^{2+}:$(CN^-)_n$ complex (as schematically shown in Fig. 11.2). This assumption explains, in a simple way, the observed redshift of the emission energy with an increase of the CN^- concentration and decrease of the excitation energy: when the CN^- concentration is increased, statistically more centers with a higher number of CN^- neighbors and lower emission energy are present in the sample, and when the excitation energy is decreased centers with larger n are excited preferentially. A similar behavior has been found for the Yb^{2+}:$(CN^-)_n$ and Eu^{2+}:$(CN^-)_n$ complexes. Differently from Yb^{2+}:$(CN^-)_n$, however, an accurate energy-level scheme as a function of n cannot be obtained from these measurements. It still can be noted, however, that the shifts of the 5d levels ($\Delta E > 100$ meV) are much more pronounced than would be expected from the purely additive behavior ($\Delta E = n \times 10$ meV) generally found for mixed-ligand cases and predicted from simple electrostatic crystal field theory [7]. This indicates that a rearrangement of the ions within the complexes takes place, leading to stronger crystal field effects. The influence of covalency and electron exchange interaction, neglected for the simple Sm^{2+}–CN^- pairs may also become more appreciable, similarly to the situation for the Yb^{2+}:$(CN^-)_n$ centers (Chap. 9). The participation of more than one Sm^{2+} ion within these complexes cannot be excluded completely. The observation that the emission is still present after quenching suggests, however, that such complexes do not play a major role. Even if this possibility is neglected, the Sm^{2+}:$(CN^-)_n$ complex configuration offers a large number of possible variations, explaining the absence of a sharp emission substructure.

11.4 E–V Energy Transfer

From the investigation of the electronic transitions of CN^--perturbed Sm^{2+} ions presented above, we can extract two types of defects which need to be distinguished:

- Simple Sm^{2+}–CN^- centers, for which the 4f-type $^5D_0$ state is the lowest excited level.
- Sm^{2+}–$(CN^-)_n$ centers with more than one molecule incorporated for which the 5d state (labeled E by Guzzi and Baldini [8]) is the lowest excited state.

The possibility of tuning the Sm^{2+} state that is most relevant to energy transfer by a change of the number n of CN^- neighbors was exploited in the work described below to study the E–V energy transfer processes.

11.4.1 Vibrational Luminescence

The question of whether E–V energy transfer takes place for any of the observed defect types can easily be answered by a measurement of the VL. In Fig. 11.3, the VL spectrum under excitation at 690 nm is shown. The signal is very weak – at least two orders of magnitude weaker than for Yb^{2+}–$(CN^-)_n$ centers [9, 10] – and consists of three main peaks. By comparison with the VL spectra from other defect systems such as $Yb^{2+}:(CN^-)_n$ and $F_H(CN^-)$ [11, 12], these can be assigned to the vibrational transitions of regular unpaired CN^- defects with different isotope compositions ($^{12}C^{14}N^-$, $^{13}C^{14}N^-$, and $^{12}C^{15}N^-$). Obviously, an efficient vibrational (V–V) energy transfer from initially excited CN^- molecules which have a Sm^{2+} neighbor to isolated CN^- molecules can take place, similarly to what has been found for $F_H(CN^-)$ centers [13] in samples with about the same CN^- concentration. The weak peak around 0.250 eV could be a candidate for VL from Sm^{2+}-perturbed CN^- molecules. For lower CN^- concentrations, for which the V-V transfer should play a minor role, the VL signal was too small to be spectrally resolved. By comparison of the integrated electronic and vibrational luminescence intensities, the quantum efficiency of the E–V transfer can be estimated to be less than 1%. In order to find out which types of $Sm^{2+}:(CN^-)_n$ complexes participate in the E–V transfer, the excitation spectrum of the VL was measured. The result for the spectral range of the dye laser used, is

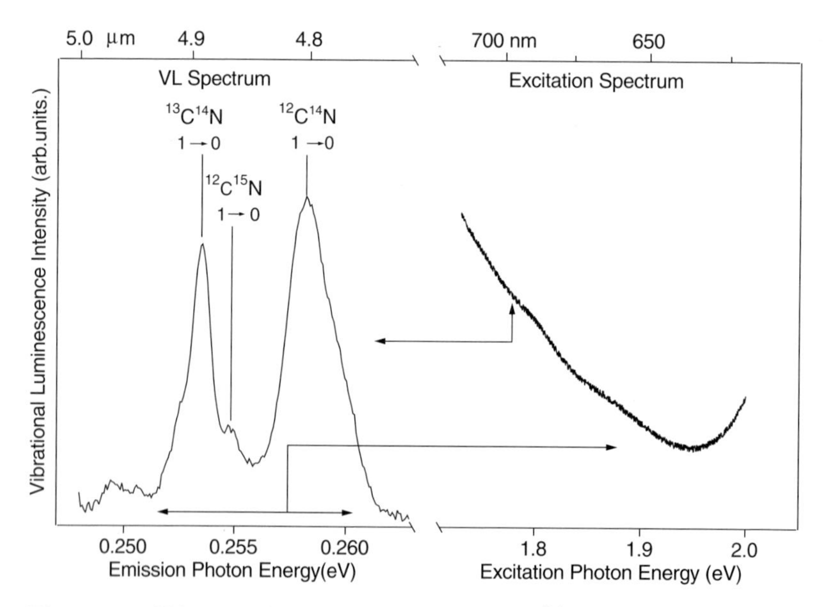

Fig. 11.3. Vibrational luminescence in KCl:Sm^{2+} + 1.5% KCN. *Left*: emission spectrum. The vibrational transitions are labeled. *Right*: excitation spectrum of VL. The spectral positions of the excitation and of probing the emission are indicated by *arrows*

illustrated in Fig. 11.3 (right) and shows that the VL can be excited over a wide spectral range. No sharp excitation peaks appear, in obvious contrast to the spectrum for the electronic emission (shown in Fig. 11.1) of the simple Sm^{2+}–CN^- sites. The VL intensity was somewhat higher in well-aged samples than in samples that had been quenched just before the measurements. This indicates that center aggregation plays a role. However, and in contrast to the case for F–CN^- and F–OH^- centers, no method for achieving a systematic pairing of the defect partners by controlled thermal annealing or light illumination could be found, despite considerable effort.

The VL signal decreases at higher temperature, similarly to what has been seen for other systems. The main cause of this behavior is the increasing nonradiative decay of the vibrational CN^- excitation. Taking this well-studied behavior of the isolated CN^- ion in KCl [14] into account, a constant or even slightly increasing E–V transfer efficiency was found up to $T = 100$ K.

11.4.2 Interpretation

The observed E–V energy transfer has a very low quantum efficiency and therefore the weak VL signals obtained do not allow a detailed study of certain aspects, such as accurate transfer rates and identification of the initially excited vibrational level. However, several clear statements can still be made:

(a) Simple Sm^{2+}–CN^- defects do not show E–V transfer as can be seen from the VL excitation spectrum. Only Sm^{2+}:$(CN^-)_n$ center complexes involving more than one CN^- molecule exhibit the transfer.
(b) The energy-level scheme of the complexes which exhibit E–V transfer is depicted in Fig. 11.4a. The complexes are characterized by 5d-type levels which lie below the 4f-type excited state and are therefore the most likely origin of the transfer.
(c) The VL excitation spectra suggest that the E–V transfer becomes more effective the more the 5d states are shifted to lower energies, i.e. the more CN^- molecules are incorporated into the defect complexes.

Several theoretical models to account for the E–V energy transfer between electronic defects and molecular vibrations have been proposed and worked out mathematically. As in Chap. 4, they can be divided into two groups: Förster–Dexter-type transfer models and models based on electronic–vibrational coupling [15, 16, 17, 18].

While the latter models are best suited to strong coupling, the FD models can easily be applied also to weakly coupled systems and to interaction over longer distances, so that they are commonly used for qualitative interpretation of the energy transfer between rare-earth ions in numerous systems. The E–V transfer models and the electron–vibrational coupling can best be visualized in the widely used configurational coordinate diagram, in which the complete vibrational-mode spectrum of the lattice with its coupling to the

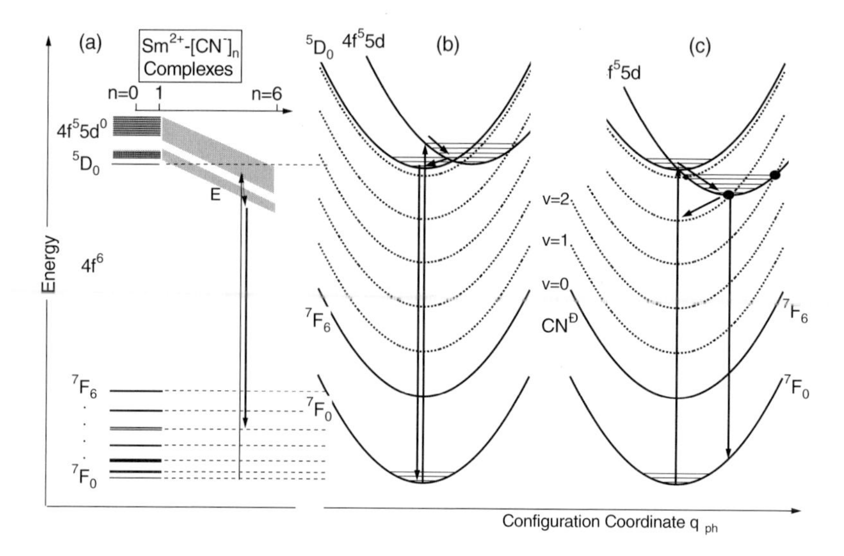

Fig. 11.4. (**a**) Simple level diagram indicating the shift of levels with increasing number n of CN^- neighbors. (**b**) and (**c**) Configurational coordinate representation of a simplified scheme of the Sm^{2+} electronic states for isolated Sm^{2+} and simple Sm^{2+}–CN^- pairs (b) and for Sm^{2+}:$(CN^-)_n$, $n > 1$ (c). For the $^7F_6$ level, vibrational CN^- energy levels are depicted as appropriately shifted additional parabolas. See text for details

electronic defects is approximated by a single mode with one displacement coordinate q_{ph} and frequency ω.

This kind of CC diagram has also been used by Fong [19] to explain the temperature dependence of the lifetime and spectral shape of the electronic luminescence of Sm^{2+} in KCl. Adopting the qualitative interpretation of the E–V transfer process in $F_H(CN^-)$ centers given by Rong et al. [20], we can make a schematic model as shown in Fig. 11.4b and c. The vibrational excitation levels are drawn as additional parabolas shifted successively upwards in energy relative to the Sm^{2+} levels by the CN^- vibrational energy (0.250 eV) for each additional v-excitation. For simplicity, this has been done only for the highest of the ground state levels ($^7F_6$) for which the most efficient E–V transfer is expected because the number of vibrational quanta required is lowest. However, the basic arguments described in the following are not influenced by this simplification. Two situations are depicted:

- In Fig. 11.4b, the weakly coupled $^5D_0$ state – represented by potential curves barely shifted relative to the $^7F_0$ ground state – is the lowest lying excited state as found for the simple Sm^{2+}–CN^- defect. In this case the excited electron relaxes rapidly into the $^5D_0$ level, and no crossing point between the electronic and vibrationally excited levels, at which a cross-over between the potential curves could take place, can be reached. This explains the absence of E–V energy transfer, at least for the low temperatures considered here.

- In Fig. 11.4c, the lowest-lying 5d state is lower in energy than the $^5D_0$ state. The 5d state exhibits a stronger coupling, reflected in a stronger horizontal shift of its potential curve. In this case, one finds several crossing points close to the potential minima at which an energy transfer is possible. This explains quite naturally the observation that E–V transfer takes place for Sm^{2+}–$(CN^-)_n$ complexes with more than one CN^- ion and low-lying 5d states.

The exact physical mechanism of the crossover from one potential curve to another depends on the model adopted. Unfortunately, the absolute energy transfer rate can be calculated within none of the available models; only relative efficiencies for the transfer channels into different CN^- vibrational states can be obtained, always with similar results: At the lowest temperatures the transfer at the lowest crossing point is the most likely. In the case depicted this would be $v = 4$.

11.4.3 Summary and Outlook

Owing to their conparatively small interaction effects and efficiency of the observed energy transfer, the Sm^{2+}:$(CN)_n$ complexes have received little attention so far. The application of the highly selective CEES technique makes – as has been shown – a distinction of the CN^--related defects from the isolated defects possible and allows a study of their properties. The results presented here are just a first step in this direction. In particular, the somewhat unique ability to tune the lowest-lying excited state by changes in the crystal field needs to be further exploited. Besides variation of the host material, application of external hydrostatic pressure could give further insights into this defect system. For KCl moderate pressures should be enough to lower the 5d states below the 4f state, which should be reflected in an increase of the E–V transfer efficiency.

References

1. J. O. Rubio, J. Phys. Chem. Solids **52**, 101 (1991).
2. F. M. M. Yasuoka, J. C. Castro, and L. A. O. Nunes, Phys. Rev. B **43**, 9295 (1991).
3. G. Baldini and M. Guzzi, Phys. Status Solidi **30**, 601 (1968).
4. W. E. Bron and W. R. Heller, Phys. Rev. **136**, 1433 (1964).
5. A. J. Ramponi and J. C. Wright, Phys. Rev. B **31**, 3965 (1985).
6. T. Tröster, private communication (unpublished).
7. H. Schläfer and G. Gliemann, *Einführung in die Ligandenfeldtheorie* (Akademische Verlagsgesellschaft, Frankfurt am Main, 1967).
8. M. Guzzi and G. Baldini, J. Lumin. **6**, 270 (1973).

9. V. Dierolf, C. P. An, and F. Luty, Abstracts for EuroDim94 (Lyon 1994).
10. M. Müller, J. F. Fabris, A. C. Hernandes, and M. Siu-Li, J. Lumin. **59**, 289 (1994).
11. F. Luty and V. Dierolf, in *Defects in Insulating Materials*, edited by O. Kanert and J.-M. Spaeth (World Scientific, Singapore, 1993), p. 17.
12. Y. Yang and F. Luty, Phys. Rev. Lett. **51**, 419 (1983).
13. V. Dierolf and F. Luty, Rev. Solid State Sci. **4**, 479 (1990).
14. K. Koch, Y. Yang, and F. Luty, Phys. Rev. B **29**, 5840 (1984).
15. G. Halama, S. H. Lin, K. T. Tsen, F. Luty, and J. B. Page, Phys. Rev. B **41**, 3136 (1990).
16. E. Gustin, M. Leblans, A. Bouwen, and D. Schoemaker, Phys. Rev. B **54**, 6977 (1996).
17. A. Naber, Ph.D. thesis, Westfälische Wilhelms-Universität Münster, 1993.
18. S. Pilzer and W. B. Fowler, Mater. Sci. Forum (Proc. ICDIM96) **239–241**, 473 (1997).
19. F. K. Fong, *Theory of Molecular Relaxation: Applications in Chemistry and Biology* (Wiley, New York, 1975).
20. F. Rong, Y. Yang, and F. Luty, Cryst. Latt. Defects and Amorph. Mater. **18**, 1 (1989).

12 Other Defect Complexes

As a third class of electronic–molecular defect complexes pairs formed by CN^- molecules and ns^2 centers, such as Tl^+ and Pb^{2+} have been studied mainly by Naber [1, 2, 3]. His essential results will be briefly summarized in this chapter.

Results from the Cu^+–OH^- defect system will also be reviewed briefly.

12.1 ns^2 Ions (Tl^+ and Pb^{2+}) and CN^- Molecules

As individual doping-induced defects in alkali halides, the Tl^+ and Pb^{2+} have been studied in great detail for a long time owing to their use as active media for spectral transformers, scintillation detectors, dosimeters, optical memory elements, tunable lasers, etc. A fairly recent review of the field which dates back to the 1920s and include the work of Pohl [4] and Hilsch [5], can be found in [6], where numerous further references are given. The optical properties of the ns^2 defects are caused by transition between the singlet $^1S_0$ electronic ground state of the ns^2 configuration to the excited states of the ns^1np^1 configurations, the lowest of which are a spin–orbit-split triplet state ($^3P_1$, $^3P_2$) and a singlet state $^1P_1$. Accordingly, three absorption bands appear in the UV, which historically are labeled A, B, and C and which correspond in the irreducible representation of the O_h group of alkali halides, to the transitions from $^1A_{1g}$ to $^3T_{1u}$ (A), $^3E_u +^3 T_{2u}$ (B), and $^1T_{1u}$ (C). While this simple (Seitz) model [7] of the ns^2 centers is able to qualitatively explain the absorption behavior, the emission can only be described when a strong Jahn–Teller interaction combined with spin–orbit coupling is taken into account. Despite this complication, several authors have tried to explain the excitation–relaxation behavior of these ions with a one-dimensional configuration coordinate diagram, as shown for Tl^+ in Fig. 12.1.

On codoping with CN^-, little change in the absorption and emission properties of the ns^2 ions are observed. Only for Pb^{2+} in KI is an extra A′ absorption band on the high-energy side of the regular A absorption found. However, no clear correlation with Pb^{2+}:$(CN^-)_n$ pairs could be established. On the contrary, there was evidence that the extra electronic absorption is due to a Pb^{2+} pair, which forms preferentially in the vicinity of the CN^-

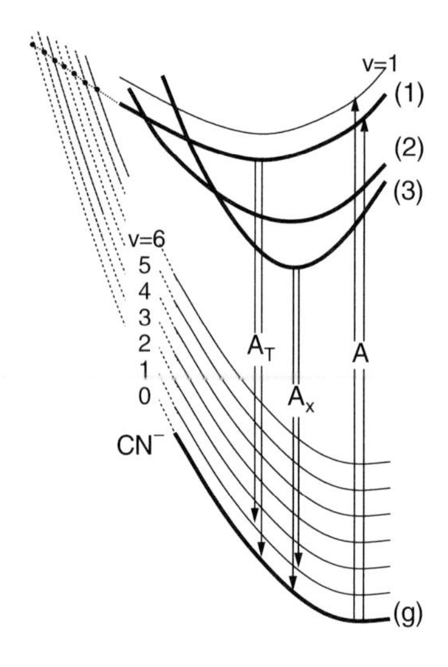

Fig. 12.1. Schematic CC diagram to illustrate the luminescence and E–V transfer properties of ns^2 centers showing (*1*) $^3T_{1u}$ level, (*2*) the $^3A_{1u}$-level, (*3*) a perturbed exciton level, and (*g*) the $^1A_{1u}$ ground state. For the latter, additional curves have been drawn which are appropriately shifted in energy and correspond to CN^- vibrational excited states (adapted from [8])

molecule. Hence the CN^- molecule causes a shift in the transitions only indirectly. Similarly, in the vibrational absorption, no energy-shifted line of perturbed CN^- molecules exists. Despite these very small static interaction effects, a VL could be found for Pb^{2+} and Tl^+ under UV excitation,[1] with an efficiency sufficiently high to cause high CN^- vibrational excitation through multiple pumping processes under intense laser irradiation. When the E–V transfer efficiency is estimated, however, a much lower value than for $F_H(CN^-)$ centers is obtained. Through time-dependent measurements, it was determined that the energy transfer leads preferentially to excitation of the CN^- molecule in its $v = 1$ or 2 level. When the vibrational excitation is depicted by extra potential curves in the CC diagram of Fig. 12.1, as done before, no intersection of the curves is found below the energy of excitation. This is due to the very different shapes of the potential curves in the ground and excited states and the large energy difference. For this reason the Förster–Dexter, sudden-approximation and supermolecule models discussed in Chap. 4 seem not to be applicable. Consequently, Naber concluded that the energy transfer takes place during the radiative transitions. Reasonably high rates can be expected for this process because the coupling parameters (the Huang–Rhys factors S_{CN} and frequency ratio β) might be substantial owing to the strong changes in the electronic environment. Examination of the absorption and emission spectra limits S_{CN} to values below 0.1 because no vibrational sidebands can be found. From these data and conclusions, it seems that the ns^2 ions and their energy transfer are quite different from what we found for the rare-earth ions and the F centers, which makes it worthwhile to study them in more detail. Despite the

[1] Naber also reported the observation of weak VL for UV-excited Sn^{2+} in CN^--doped KBr and Cu^+ in KCl.

work of Naber, several questions remain unanswered, most notably about the type of defects which participate in the transfer. As we have seen for the rare-earth ions, this may not be the simple ns^2 ion–CN^- pairs, but more complicated clusters. Unfortunately, excitation has to be performed in the UV, for which no (easy to handle) tunable lasers are presently available to perform excitation spectroscopy which would be decisive for clarification of this point.

12.2 Cu^+ Ions and OH^- Molecules

In NaCl crystals doped with Cu^+ ions in the melt or by diffusion a blue emission at around 420 nm was found, which was not expected for the $4d^{10} \leftrightarrow 4d^9s^1$ transitions of isolated Cu^+ defects in alkali halides. At first this new emission was attributed to Cu^+ complexes [9]. Only later did a systematic investigation with more carefully prepared samples show that the emission is connected to the presence of unwanted OH^- impurities in the NaCl [10, 11]. After this clarification of the origin of the emission, more studies of the emission and absorption properties were carried out. These revealed rather drastic changes in the spectral positions and radiative lifetimes of the Cu^+ ion when perturbed by an OH^- neighbor [12]. The vibrational properties of such a defect system were studied by Fabris et al. in NaF [13]. These authors found a large number of VA lines redshifted with respect to the regular OH^- lines which exhibited slightly enhanced mechanical anharmonicities and a wide variation in their electric-anharmonicity values. A clear identification of the origin of the various lines has not been achieved yet, but it was speculated that they were related to different relative arrangements of the centers. The observed spectral positions and corresponding shifts of the electronic luminescence and vibrational absorption indicated that rather strong mutual interaction effects are present, which may lead to other phenomena (E–V transfer, bistability, and enhancement of transition probabilities) as well. However, no studies have been performed in this direction so far.

References

1. A. Naber and F. Fischer, Radiat. Eff. Defects Solids **119–121**, 553 (1993).
2. A. Naber, in *Defects in Insulating Materials*, edited by O. Kanert and J.-M. Spaeth (World Scientific, Singapore, 1993), p. 543.
3. A. Naber, Ph.D. thesis, Westfälische Wilhelms-Universität Münster, 1993.
4. R. Pohl, Proc. Phys. Soc. **49**, 3 (1937).
5. R. Hilsch, Proc. Phys. Soc. **49**, 40 (1937).
6. S. Zazubovich, Int. J. of Mod. Phys. **8**, 985 (1994).
7. F. Seitz, J. Chem. Phys. **6**, 150 (1938).

8. R. Illingworth, Phys. Rev. **136**, A504 (1965).
9. T. Kurobori, S. Taniguchi, and N. Takeuchi, J. Mat. Sci. Lett. **11**, 1140 (1992).
10. T. Kurobori, S. Taniguchi, and N. Takeuchi, Phys. Status. Solidi B **172**, K77 (1992).
11. T. Kurobori, S. Taniguchi, and N. Takeuchi, J. Lumin **55**, 183 (1993).
12. T. Kurobori, H. Yonezawa, and N. Takeuchi, J. Lumin. **59**, 157 (1994).
13. J. Fabris, M. Müller, A. Hernandes, and M. Siu-Li, Radiat. Eff. Defects Solids **133**, 321 (1995).

13 Summary

In this final chapter, we compare the defect systems that we have discussed and summarize their properties in a tablc. Thc implications for their use as models systems and for their applications will be discussed as well.

13.1 Comparison of the Defect Systems

In the course of this text, the following interaction phenomena between electronic or monatomic defects on one hand and diatomic molecules on the other have been described:

- shifts of electronic transition energies,
- changes of vibrational transition energies and anharmonicities,
- changes of vibrational transition probabilities,
- enhancement of radiative transition probabilities due to dynamic interaction,
- luminescence quenching,
- optical and thermally induced bistability,
- electronic–vibrational coupling and sidebands,
- E–V energy transfer.

The number of effects and the discovery of new properties made it well worth investigating these defect systems in detail, although, as pointed out in the introduction, the initial goal of increasing the radiative lifetime of the F luminescence was not fulfilled. Only many years later could it be shown, for a different complex (Yb^{2+}–CN^{-}), that molecular defect partners can indeed have the desired effect. The discussion of most of the properties has already been carried out for the various defect systems individually in the respective chapters, and therefore the focus will be here on comparative aspects. For that purpose, the main properties are summarized in Table 13.1 along with a qualitative judgment of how prominent each effect is for the respective defect system.

As can be seen from Table 13.1, the experimental characterization is quite complete for most systems. Many of the gaps ("n/o") are just a consequence of the fact that the particular effect is very small.

Table 13.1. Comparison of interaction effects for different defect systems and host materials. The following symbols are used: =, no change ; +, small value or intermediate change in positive direction; ++, large value, strong changes ; – and – –; same for negative direction; 0, very small value; n/o, not observed; Δ, splitting; EA,VA, positions of EA and VA; qu, EL is quenched. The interaction phenomena are characterized by Q_{VA}, integrated VA strength; x_{e}, mechanical anharmonicity parameter; f_{e}, oscillator strength of electronic transitions; I_{EL}, intensity of EL, S_{ph}; S_{CN}, Huang–Rhys factors; BS, bistability; v_{dest}, v-level populated through E–V transfer. X stands for Cl, Br, I

Changes in		EA	VA	Q_{VA}	x_{e}	f_{e}	I_{El}	S_{ph}	S_{CN}	BS	E–V transfer	
System	Host										$W_{\mathrm{E-V}}$	v_{dest}
$F–CN^-$	KX, RbX	– (shift)	0	0	0	=	=	++	0	no	+	3.4,5
	CsX	+ (Δ)	–	+	+	=	qu	++	+	no	++	3,4,5
$F'–CN^-$	CsBr,KCl	=	–	++	n/o	n/o	n/o	++	n/o	n/o	n/o	n/o
$F–OH^-$	K, Rb	± (BS)	∓	=	=	=	=	++	n/o	yes	+	1,2
	Cs	++ (Δ)	–	+	+	=	qu	++	n/o	no	+	1,2
$(OH^-)^-$	CsX	–	– –	++	++	=	n/o	++	n/o	n/o	+	1,2
$Yb^{2+}–CN^-$	KX,NaCl	–	+	+	n/o	+	+	+	0	yes	0	?
$Yb^{2+}:(CN^-)_n$	KX,NaCl	– –	+	+	n/o	++	+ to qu	+	0	no	++	6,7
$Eu^{2+}:(CN^-)_n$	KX	–	+	n/o	n/o	n/o	+	+	n/o	n/o	+	?
$Sm^{2+}–CN^-$	KCl	4f: 0; 5d:±,	n/o	n/o	n/o	=	=	0,+	n/o	n/o	n/o	n/o
$Sm^{2+}–(CN^-)_n$	KCl	5d: – –	n/o	n/o	n/o	=	+	+	n/o	n/o	+	?
$ns^2–CN^-$	KX	n/o	=	n/o	n/o	n/o	n/o	++	n/o	n/o	0	1,2

Although the details are complex and cannot be included in the table, several general trends can be seen. For instance, the “interaction strength” is reflected at least qualitatively in all of the effects. In other words, systems with large spectral shifts also exhibit a pronounced E–V energy transfer, and so on. However, the finer details of the properties also play a significant role. The coupling of the electronic states to the phonons drastically influences the E–V transfer efficiency. This could be shown most clearly for the $Sm^{2+}–CN^-$system in KCl but also becomes apparent in a comparison between different defect systems. Furthermore, an “energy gap law” seems to apply and is well justified theoretically, independently of the particular model used. This rule states that the E–V transfer is more likely if the amount of energy which needs to be transferred is small.

Comparing the systems with respect to their E–V transfer properties, at least four different mechanisms have to be distinguished:

(a) Transfer originating from a relaxed excited state into a vibrational level which corresponds in total energy approximately to the energy of the competing electronic-luminescence process. This type of mechanism has been encountered in $F_H(CN^-)$ centers and Yb^{2+}:$(CN^-)_n$ complexes.
(b) Transfer during configurational relaxation into vibrational levels $v = 1, 2$, as seen for $F_H(OH^-)$ centers.
(c) Transfer accompanied by electron tunneling, as seen between F centers and OH^- defects which are several lattice parameters apart. This process results in molecular excitation into $v = 1, 2$.
(d) Transfer during electronic transitions claimed for the ns^2-type defects, for which all other processes are highly unlikely.

Although this classification seems to indicate that different physical mechanisms are at work, it might also just be the consequence of different interaction strengths. In particular, the first two processes may be distinguished only by the transfer efficiency. While transfer efficiency may be high enough to compete with the fast relaxation in some defect systems, in others it may be much lower allowing only an efficient competition with the EL in the comparatively long-lived RES. Although it was possible to achieve a good phenomenological understanding of the E–V transfer, the physics of the E–V transfer is still not completely clarified, in the author's opinion. While at the moment it seems that there are only a few experimental measurements that could be done to improve our understanding of the systems, it is apparent that many questions still need to be answered from the theoretical point of view.

In the theoretical studies of the electron–molecular interaction effects which have been performed so far, several phenomenological models have been applied but in many cases their limitations have become obvious. There are several reasons for this:

- The models were based on interacting but otherwise independent defects, such that as soon as the mutual perturbation effects become strong this approach must fail.
- On the other hand, even for isolated defects, the models have their restrictions. In the course of this text, this became apparent from the difficulties in describing the RES of F centers, predicting the electric dipole moments of isolated OH^- defects, and qualitatively describing covalency effects in the electronic absorption of Yb^{2+} ions in alkali halides. In view of this lack of understanding of the individual defects, it is not surprising that a quantitative description of the defect complexes is not possible on this basis.

- The agreement of the standard theories for individual defects is in many cases only qualitative, and only general tendencies can be predicted. The same holds for the complexes. In both cases the quantitative agreement which could be achieved for some systems may be accidental.

It can be seen that many open questions remain, and I would like to encourage theorists to tackle the description of these interesting defect systems with a combination of methods from solid-state physics and molecular quantum chemistry. Along with detailed experimental characterization,[1] this could yield a solid basis which could then be generalized and applied to more complicated systems that might have potential important applications in the future.

13.2 Potential Applications

While the role of the defect complexes as "model" discussed above systems is not a matter of debate, potential commercial applications are still far away. Although some possibilities, such as LEDs and lasers, have been pointed out and their feasibility has been demonstrated, there are practical obstacles such as the limitation to low temperatures or the toxic nature of CN^-. Furthermore, the mechanical and thermal properties of alkali halides are rather unfavorable. These restrictions make it virtually impossible that the prototypes described above will become widespread, everyday products. Nevertheless, the concepts can be generalized to other materials and other electronic/molecular defect systems. Devices such as lasers, dosimeters, colored LEDs, photorefractive memory, and X-ray storage phosphors may be based on the ideas developed. The drastic influence of molecules on the properties of other defects and vice versa is limited neither to CN^- and OH^- nor to alkali halide crystals. On the contrary, these restrictions have been chosen here just for the sake of simplicity. In particular, it is worthwhile to scan through those defect systems in ionic materials which already have established applications, and to reexamine the question whether they can be optimized or customized for other uses by the introduction of molecular defect partners. Besides this positive influence of the molecules on the properties of defects, a better understanding of the performance-limiting processes often induced by "unwanted" molecular impurities may ultimately lead to practical solutions other than the straightforward but often impossible one of avoiding those impurities. The OH^- molecule in oxide and fluoride laser materials is the most widespread example in the latter category and is often quoted as the culprit in cases of bad device performance.

[1] Sometimes, however, experimental information that is too detailed may discourage the application of theoretical models because the likelihood that they can be ruled out is much higher, and the effort then appears to be in vain [1].

The interaction effects observed between the defect partners in alkali halides are, in the case of the rare-earth–CN^- complexes, accompanied by similar changes in the melt and in solid solution. This observation suggests that the special properties of these complexes are not limited to the crystalline phase, and encourages the investigation of the these defect systems in glasses which, compared with crystalline hosts, offer much more flexibility in terms of the formation of complexes, fabrication, and potential applications.

Reference

1. P. W. Gash, private communication.

Index

Printing: Mercedes-Druck, Berlin
Binding: Stein+Lehmann, Berlin